Modern Chemistry®

Datasheets for In-Text Labs with Lab Notes and Answer Key

HOLT, RINEHART AND WINSTON

A Harcourt Education Company

Orlando • **Austin** • New York • San Diego • Toronto • London

Printed in the United States of America

ISBN 0-03-037112-0

2 3 4 5 6 7 8 9   170   07 06 05

Contents

Ordering Lab Materials with the
One-Stop Planner® CD-ROM

Your class and prep time are valuable. Now, it's easier and faster than ever to organize and obtain the materials that you need for all of the labs in the *Modern Chemistry* lab program. Using the Lab Materials QuickList software found on the *One-Stop Planner® CD-ROM*, you can do the following:

• View all of the materials that you need for any (or all) labs.
The Lab Materials QuickList software allows you to easily see all of the materials needed for any lab in the *Modern Chemistry* lab program. Use your materials list to order all of your materials at once. Or use the list to determine what items you need to resupply or supplement your stockroom so that you'll be prepared to do any lab.

• Create a customized materials list.
No two teachers teach exactly alike. The Lab Materials QuickList software allows you to select labs by type of lab or by type of material used. You can create a materials list that summarizes the needs of whichever labs from *Modern Chemistry* you choose.

• Let the software handle the details.
You can customize your list based on the number of students and the number of lab groups. A powerful software engine that has been programmed to distinguish between consumable and nonconsumable materials will "do the math." Whether you're examining all of the labs for a whole year or just the labs that you're planning for next week, the software does the hard work of totaling and tallying, telling you what you need and *exactly* how much you'll need for the labs that you've selected.

• Print your list.
By printing out materials lists that you created by using the Lab Materials QuickList software, you can have a copy of any materials list right at your fingertips for easy reference at any time.

• Order your materials easily.
After you've created your materials list by using the Lab Materials QuickList software, you can use it to order from Sargent-Welch or to prepare a purchase order to be sent directly to another scientific materials supplier.

Visit go.hrw.com to learn more about the Lab Materials QuickList software.

Teacher's Laboratory Safety Information

HELPING STUDENTS RECOGNIZE THE IMPORTANCE OF SAFETY

One method that can help students appreciate the importance of precautions is to use a safety contract that students read and sign, indicating they have read, understand, and will respect the necessary safety procedures, as well as any other written or verbal instructions that will be given in class. You can find a copy of a safety contract on the *Modern Chemistry One Stop Planner CD-ROM.* You can use this form as a model or make your own safety contract for your students with language specific to your lab situation. When making your own contract, you could include points such as the following:

- Make sure that students agree to wear personal protective equipment (goggles and lab aprons) at all times. Safety information regarding the use of contact lenses is continually changing. Check your state and local regulations on this subject. Students should agree to read all lab exercises before they come to class. They should agree to follow all directions and safety precautions and to use only materials and equipment that you provide.
- Students should agree to remain alert and cautious at all times in the lab. They should never leave experiments unattended.
- Students should not wear dangling jewelry or bulky clothing.
- Students should bring only lab manuals and lab notebooks into the lab. Backpacks, textbooks for other subjects, and other items should be stored elsewhere.
- Students should agree to never eat, drink, or smoke in any science laboratory. Food should **never** be brought into the laboratory
- Students should **never** taste or touch chemicals.
- Students should keep themselves and other objects away from Bunsen burner flames. Students should be responsible for checking, before they leave, that gas valves and hot plates are off.
- Students should know the proper fire drill procedures and the locations of fire exits.
- Students should always clean all apparatus and work areas.
- Students should wash their hands thoroughly with soap and water before leaving the lab room.
- Students should know the locations and operation of all safety equipment in the laboratory.
- Students should report all accidents or close calls to you immediately, no matter how minor.
- Students should **never** work alone in the laboratory, and they should never work unless you are present.

DISPOSAL OF CHEMICALS

Only a relatively small percentage of waste chemicals are classified as hazardous by EPA regulations. The EPA regulations are derived from two acts (as amended) passed by the Congress of the United States: RCRA (Resource Conservation and Recovery Act) and CERCLA (Comprehensive Environmental Response, Compensation, and Liability Act).

In addition, some states have enacted legislation governing the disposal of hazardous wastes that differs to some extent from the federal legislation. The disposal procedures described in this book have been designed to comply with the federal legislation as described in the EPA regulations.

In most cases the disposal procedures indicated in the teacher's edition will probably comply with your state's disposal requirements. However, to be sure of this, check with your state's environmental agency. If a particular disposal procedure does not comply with your state requirements, ask that agency to assist you in devising a procedure that is in compliance.

The following general practices are recommended in addition to the specific instructions given in the Lab Notes.

- Except when otherwise specified in the disposal procedures, neutralize acidic and basic wastes with 1.0 M potassium hydroxide, KOH, or 1.0 M sulfuric acid, H_2SO_4, added slowly and stirred.

- In dealing with a waste disposal contractor, prepare a complete list of the chemicals you want to dispose of. Classify each chemical on your disposal list as hazardous or nonhazardous waste. Check with your local environmental agency office for the details of such classification.

- Unlabeled bottles are a special problem. They must be identified to the extent that they can be classified as hazardous or nonhazardous waste. Some landfills will analyze a mystery bottle for a fee if it is shipped to the landfill in a separate package, is labeled as a sample, and includes instructions to analyze the contents sufficiently to allow proper disposal.

ELECTRIC SAFETY

Although none of the labs in this manual require electrical equipment, several include options for the use of microcomputer-based laboratory equipment, pH meters, or other equipment. The following safety precautions to avoid electrical shocks must be observed any time electrical equipment is present in the lab:

- Each electrical socket in the laboratory must be a three-hole socket and must be protected with a GFI (ground-fault interrupter) circuit.

- Check the polarity of all circuits with a polarity tester from an electronics supply store before you use them. Repair any incorrectly wired sockets.

- Use only electrical equipment that has a three-wire cord and three-prong plug.

- Be sure all electrical equipment is turned off before it is plugged into a socket. Turn off electrical equipment before you unplug it.

- Wiring hookups should be made or altered only when apparatus is disconnected from the power source and the power switch is turned off.

- Do not let electrical cords dangle from work stations; dangling cords are a shock hazard, and students can trip over them.
- Do not use electrical equipment with frayed or twisted cords.
- The area under and around electrical equipment should be dry; cords should not lie in puddles of spilled liquid.
- Your hands should be dry when you use electrical equipment.
- Do not use electrical equipment powered by 110–115 V alternating current for conductivity demonstrations or for any other use in which bare wires are exposed, even if the current is connected to a lower-voltage AC or DC connection.

Use dry cells or nicad rechargeable batteries as direct-current sources. Do not use automobile storage batteries or AC-to-DC converters; these two sources of DC current can present serious shock hazards.

Prepared by Jay A. Young, Consultant, Chemical Health and Safety, Silver Spring, Maryland

Incompatible Chemicals

Consider the following list when organizing and storing chemicals. Note that some chemicals on this list should no longer be in your lab, due to their potential risks. Consult your state and local guidelines for more specific information on chemical hazard.

Chemical	Should not come in contact with
acetic acid	chromic acid, nitric acid, perchloric acid, ethylene glycol, hydroxyl compounds, peroxides, or permanganates
acetone	concentrated sulfuric acid or nitric acid mixtures
acetylene	silver, mercury, or their compounds; bromine, chlorine, fluorine, or copper tubing
alkali metals, powdered aluminum or magnesium	water, carbon dioxide, carbon tetrachloride, or the halogens
ammonia (anhydrous)	mercury, hydrogen fluoride, or calcium hypochlorite
ammonium nitrate (strong oxidizer)	strong acids, metal powders, chlorates, sulfur, flammable liquids, or finely divided organic materials
aniline	nitric acid or hydrogen peroxide
bromine	ammonia, acetylene, butane, hydrogen, sodium carbide, turpentine, or finely divided metals
carbon (activated)	calcium hypochlorite or any oxidizing agent
chlorates	ammonium salts, strong acids, powdered metals, sulfur, or finely divided organic materials
chromic acid	glacial acetic acid, camphor, glycerin, naphthalene, turpentine, low-molar-mass alcohols, or flammable liquids
chlorine	same as bromine
copper	acetylene or hydrogen peroxide
flammable liquids	ammonium nitrate, chromic acid, hydrogen peroxide, sodium peroxide, nitric acid, or any of the halogens
hydrocarbons (butane, propane, gasoline, turpentine)	fluorine, chlorine, bromine, chromic acid, or sodium peroxide
hydrofluoric acid	ammonia
hydrogen peroxide	most metals or their salts, flammable liquids, or other combustible materials
hydrogen sulfide	nitric acid or certain other oxidizing gases
iodine	acetylene or ammonia
nitric acid	glacial acetic acid, chromic or hydrocyanic acids, hydrogen sulfide, flammable liquids, or flammable gases that are easily nitrated

Chemical	Should not come in contact with
oxygen	oils, grease, hydrogen, or flammable substances
perchloric acid	acetic anhydride, bismuth or its alloys, alcohols, paper, wood, or other organic materials
phosphorus pentoxide	water
potassium permanganate	glycerin, ethylene glycol, or sulfuric acid
silver	acetylene, ammonium compounds, oxalic acid, or tartaric acid
sodium peroxide	glacial acetic acid, acetic anhydride, methanol, carbon disulfide, glycerin, benzaldehyde, or water
sulfuric acid	chlorates, perchlorates, permanganates, or water

Introduction to the Lab Program

Structure of the Experiments

INTRODUCTION

The opening paragraphs set the theme for the experiment and summarize its major concepts.

OBJECTIVES

Objectives highlight the key concepts to be learned in the experiment and emphasize the science process skills and techniques of scientific inquiry.

MATERIALS

These lists enable you to organize all apparatus and materials needed to perform the experiment. Knowing the concentrations of solutions is vital. You often need this information to perform calculations and to answer the questions at the end of the experiment.

SAFETY

Safety cautions are placed at the beginning of the experiment to alert you to procedures that may require special care. Before you begin, you should review the safety issues that apply to the experiment.

PROCEDURE

By following the procedures of an experiment, you perform concrete laboratory operations that duplicate the fact-gathering techniques used by professional chemists. You learn skills in the laboratory. The procedures tell you how and where to record observations and data.

DATA AND CALCULATIONS TABLES

The data that you collect during each experiment should be recorded in the labeled Data Tables provided. The entries you make in a Calculations Table emphasize the mathematical, physical, and chemical relationships that exist among the accumulated data. Both types of tables should help you to think logically and to formulate your conclusions about what occurs during the experiment.

CALCULATIONS

Space is provided for all computations based on the data you gather.

QUESTIONS

Based on the data and calculations, you should be able to develop plausible explanations for the phenomena you observe during the experiment. Specific questions require you to draw on the concepts you learn.

GENERAL CONCLUSIONS

This section asks broader questions that bring together the results and conclusions of the experiment and relate them to other situations.

Safety in the Chemistry Laboratory

CHEMICALS ARE NOT TOYS.

Any chemical can be dangerous if it is misused. Always follow the instructions for the experiment. Pay close attention to the safety notes. Do not do anything differently unless told to do so by your teacher.

Chemicals, even water, can cause harm. The trick is to know how to use chemicals correctly so that they will not cause harm. You can do this by following the rules on these pages, paying attention to your teacher's directions, and following the cautions on chemical labels and experiments.

These safety rules always apply in the lab.

1. **Always wear a lab apron and safety goggles.**
   Even if you aren't working on an experiment, laboratories contain chemicals that can damage your clothing, so wear your apron and keep the strings of the apron tied. Because chemicals can cause eye damage, even blindness, you must wear safety goggles. If your safety goggles are uncomfortable or get clouded up, ask your teacher for help. Try lengthening the strap a bit, washing the goggles with soap and warm water, or using an antifog spray.

2. **Generally, no contact lenses are allowed in the lab.**
   Even while wearing safety goggles, you can get chemicals between contact lenses and your eyes, and they can cause irreparable eye damage. If your doctor requires that you wear contact lenses instead of glasses, then you may need to wear special safety goggles in the lab. Ask your doctor or your teacher about them.

3. **Never work alone in the laboratory.**
   You should always do lab work under the supervision of your teacher.

4. **Wear the right clothing for lab work.**
   Necklaces, neckties, dangling jewelry, long hair, and loose clothing can cause you to knock things over or catch items on fire. Tuck in a necktie or take it off. Do not wear a necklace or other dangling jewelry, including hanging earrings. It isn't necessary, but it might be a good idea to remove your wristwatch so that it is not damaged by a chemical splash.

   Pull back long hair, and tie it in place. Nylon and polyester fabrics burn and melt more readily than cotton, so wear cotton clothing if you can. It's best to wear fitted garments, but if your clothing is loose or baggy, tuck it in or tie it back so that it does not get in the way or catch on fire.

   Wear shoes that will protect your feet from chemical spills—no open-toed shoes, sandals, or shoes made of woven leather straps. Shoes made of solid leather or a polymer are much better than shoes made of cloth. Also, wear pants, not shorts or skirts.

5. **Only books and notebooks needed for the experiment should be in the lab.**
   Do not bring other textbooks, purses, bookbags, backpacks, or other items into the lab; keep these things in your desk or locker.

6. Read the entire experiment before entering the lab.
Memorize the safety precautions. Be familiar with the instructions for the experiment. Only materials and equipment authorized by your teacher should be used. When you do the lab work, follow the instructions and the safety precautions described in the directions for the experiment.

7. Read chemical labels.
Follow the instructions and safety precautions stated on the labels. Know the location of Material Safety Data Sheets for chemicals.

8. Walk carefully in the lab.
Sometimes you will carry chemicals from the supply station to your lab station. Avoid bumping other students and spilling the chemicals. Stay at your lab station at other times.

9. Food, beverages, chewing gum, cosmetics, and tobacco are *never* allowed in the lab.
You already know this.

10. Never taste chemicals or touch them with your bare hands.
Also, keep your hands away from your face and mouth while working, even if you are wearing gloves.

11. Use a sparker to light a Bunsen burner.
Do not use matches. Be sure that all gas valves are turned off and that all hot plates are turned off and unplugged before you leave the lab.

12. Be careful with hot plates, Bunsen burners, and other heat sources.
Keep your body and clothing away from flames. Do not touch a hot plate just after it has been turned off. It is probably hotter than you think. Use tongs to heat glassware, crucibles, and other things and to remove them from a hot plate, a drying oven, or the flame of a Bunsen burner.

13. Do not use electrical equipment with frayed or twisted cords or wires.

14. Be sure your hands are dry before you use electrical equipment.
Before plugging an electrical cord into a socket, be sure the electrical equipment is turned off. When you are finished with it, turn it off. Before you leave the lab, unplug it, but be sure to turn it off first.

15. Do not let electrical cords dangle from work stations; dangling cords can cause tripping or electric shocks.
The area under and around electrical equipment should be dry; cords should not lie in puddles of spilled liquid.

16. Know fire drill procedures and the locations of exits.

17. Know the locations and operation of safety showers and eyewash stations.

18. If your clothes catch on fire, *walk* to the safety shower, stand under it, and turn it on.

19. **If you get a chemical in your eyes, walk immediately to the eyewash station, turn it on, and lower your head so that your eyes are in the running water.**

Hold your eyelids open with your thumbs and fingers, and roll your eyeballs around. You have to flush your eyes continuously for at least 15 min. Call your teacher while you are doing this.

20. **If you have a spill on the floor or lab bench, don't try to clean it up by yourself.**

First, ask your teacher if it is OK for you to do the cleanup; if it is not, your teacher will know how the spill should be cleaned up safely.

21. **If you spill a chemical on your skin, wash it off under the sink faucet, and call your teacher.**

If you spill a solid chemical on your clothing, brush it off carefully so that you do not scatter it, and call your teacher. If you get a liquid chemical on your clothing, wash it off right away if you can get it under the sink faucet, and call your teacher. If the spill is on clothing that will not fit under the sink faucet, use the safety shower. Remove the affected clothing while under the shower, and call your teacher. (It may be temporarily embarrassing to remove your clothing in front of your class, but failing to flush that chemical off your skin could cause permanent damage.)

22. **The best way to prevent an accident is to stop it before it happens.**

If you have a close call, tell your teacher so that you and your teacher can find a way to prevent it from happening again. Otherwise, the next time, it could be a harmful accident instead of just a close call.

23. **All accidents should be reported to your teacher, no matter how minor.**

Also, if you get a headache, feel sick to your stomach, or feel dizzy, tell your teacher immediately.

24. **For all chemicals, take only what you need.**

On the other hand, if you do happen to take too much and have some left over, **do not** put it back in the bottle. If somebody accidentally puts a chemical into the wrong bottle, the next person to use it will have a contaminated sample. Ask your teacher what to do with any leftover chemicals.

25. *Never* **take any chemicals out of the lab.**

You should already know this rule.

26. **Horseplay and fooling around in the lab are very dangerous.**

Never be a clown in the laboratory.

27. **Keep your work area clean and tidy.**

After your work is done, clean your work area and all equipment.

28. **Always wash your hands with soap and water before you leave the lab.**

29. **Whether or not the lab instructions remind you,** *all* **of these rules** *apply all of the time.*

QUIZ

Determine which safety rules apply to the following.

- Tie back long hair, and confine loose clothing. (Rule ? applies.)

- Never reach across an open flame. (Rule ? applies.)

- Use proper procedures when lighting Bunsen burners. Turn off hot plates and Bunsen burners that are not in use. (Rule ? applies.)

- Be familiar with the procedures and know the safety precautions before you begin. (Rule ? applies.)

- Use tongs when heating containers. Never hold or touch containers with your hands while heating them. Always allow heated materials to cool before handling them. (Rule ? applies.)

- Turn off gas valves that are not in use. (Rule ? applies.)

SAFETY SYMBOLS

To highlight specific types of precautions, the following symbols are used in the experiments. Remember that no matter what safety symbols and instructions appear in each experiment, all of the 29 safety rules described previously should be followed at all times.

EYE AND CLOTHING PROTECTION

- Wear safety goggles in the laboratory at all times. Know how to use the eyewash station.

- Wear laboratory aprons in the laboratory. Keep the apron strings tied so that they do not dangle.

CHEMICAL SAFETY

- Never taste, eat, or swallow any chemicals in the laboratory. Do not eat or drink any food from laboratory containers. Beakers are not cups, and evaporating dishes are not bowls.

- Never return unused chemicals to their original containers.

- Some chemicals are harmful to the environment. You can help protect the environment by following the instructions for proper disposal.

- It helps to label the beakers and test tubes containing chemicals.

- Never transfer substances by sucking on a pipet or straw; use a suction bulb.

- Never place glassware, containers of chemicals, or anything else near the edges of a lab bench or table.

CAUSTIC SUBSTANCES

- If a chemical gets on your skin or clothing or in your eyes, rinse it immediately, and alert your teacher.

- If a chemical is spilled on the floor or lab bench, tell your teacher, but do not clean it up yourself unless your teacher says it is OK to do so.

HEATING SAFETY

- When heating a chemical in a test tube, always point the open end of the test tube away from yourself and other people.

EXPLOSION PRECAUTION

- Use flammable liquids in small amounts only.
- When working with flammable liquids, be sure that no one else in the lab is using a lit Bunsen burner or plans to use one. Make sure there are no other heat sources present.

HAND SAFETY

- Always wear gloves or use cloths to protect your hands when cutting, fire polishing, or bending hot glass tubing. Keep cloths clear of any flames.
- Never force glass tubing into rubber tubing, rubber stoppers, or corks. To protect your hands, wear heavy leather gloves or wrap toweling around the glass and the tubing, stopper, or cork, and gently push the glass tubing into the rubber or cork.
- Use tongs when heating test tubes. Never hold a test tube in your hand to heat it.
- Always allow hot glassware to cool before you handle it.

GLASSWARE SAFETY

- Check the condition of glassware before and after using it. Inform your teacher of any broken, chipped, or cracked glassware because it should not be used.
- Do not pick up broken glass with your bare hands. Place broken glass in a specially designated disposal container.

GAS PRECAUTION

- Do not inhale fumes directly. When instructed to smell a substance, waft it toward you. That is, use your hand to wave the fumes toward your nose. Inhale gently.

RADIATION PRECAUTION

- Always wear gloves when handling a radioactive source.
- Always wear safety goggles when performing experiments with radioactive materials.
- Always wash your hands and arms thoroughly after working with radioactive materials.

HYGIENIC CARE

- Keep your hands away from your face and mouth.
- Always wash your hands before leaving the laboratory.

Any time you see any of the safety symbols, you should remember that all 29 of the numbered laboratory rules always apply.

Labeling of Chemicals

In any science laboratory the labeling of chemical containers, reagent bottles, and equipment is essential for safe operations. Proper labeling can lower the potential for accidents that occur as a result of misuse. Read labels and equipment instructions several times before you use chemicals or equipment. Be sure that you are using the correct items, that you know how to use them, and that you are aware of any hazards or precautions associated with their use.

All chemical containers and reagent bottles should be labeled prominently and accurately with labeling materials that are not affected by chemicals.

Chemical labels should contain the following information:

1. **Name of the chemical and its chemical formula**

2. **Statement of possible hazards** This is indicated by the use of an appropriate signal word, such as *DANGER*, *WARNING*, or *CAUTION*. This signal word usually is accompanied by a word that indicates the type of hazard present, such as *POISON*, *CAUSES BURNS*, *EXPLOSIVE*, or *FLAMMABLE*. Note that this labeling should not take the place of reading the appropriate Material Safety Data Sheet for a chemical.

3. **Precautionary measures** Precautionary measures describe how users can avoid injury from the hazards listed on the label. Examples include: "use only with adequate ventilation" and "do not get in eyes or on skin or clothing."

4. **Instructions in case of contact or exposure** If accidental contact or exposure does occur, immediate first-aid measures can minimize injury. For example, the label on a bottle of acid should include this instruction: "In case of contact, flush with large amounts of water; for eyes, rinse freely with water for 15 min and get medical attention immediately."

5. **The date of preparation and the name of the person who prepared the chemical** This information is important for maintaining a safe chemical inventory.

Suggested Labeling Scheme	
Name of contents	hydrochloric acid
Chemical formula and concentration or physical state	6 M HCl
Statements of possible hazards and precautionary and measures	WARNING! CAUSTIC and CORROSIVE—CAUSES BURNS Avoid contact with skin and eyes Avoid breathing vapors.
Hazard Instructions for contact or overexposure	IN CASE OF CONTACT: Immediately flush skin or eyes with large amounts of water for at least 15 min; for eyes, get medical attention immediately!
Date prepared or obtained Manufacturer (commercially obtained) or preparer (locally made)	May 8, 2005 Prepared by Betsy Byron, Faribault High School, Faribault, Minnesota

Laboratory Techniques

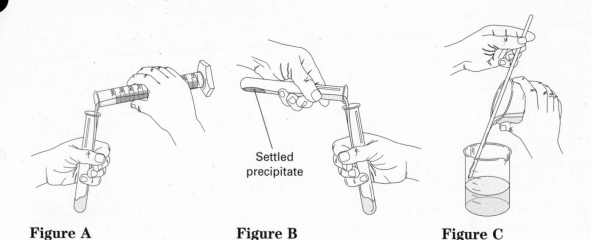

Figure A   Figure B   Figure C

Settled precipitate

DECANTING AND TRANSFERRING LIQUIDS

1. The safest way to transfer a liquid from a graduated cylinder to a test tube is shown in **Figure A.** Transfer the liquid at arm's length with your elbows slightly bent. This position enables you to see what you are doing and still maintain steady control.

2. Sometimes liquids contain particles of insoluble solids that sink to the bottom of a test tube or beaker. Use one of the methods given below to separate a supernatant (the clear fluid) from insoluble solids.

   a. Figure B shows the proper method of decanting a supernatant liquid in a test tube.

   b. Figure C shows the proper method of decanting a supernatant liquid in a beaker by using a stirring rod. The rod should touch the wall of the receiving container. Hold the stirring rod against the lip of the beaker containing the supernatant liquid. As you pour, the liquid will run down the rod and fall into the beaker resting below. Using this method will prevent the liquid from running down the side of the beaker you are pouring from.

HEATING SUBSTANCES AND EVAPORATING SOLUTIONS

1. Use care in selecting glassware for high-temperature heating. The glassware should be heat resistant.

2. When using a gas flame to heat glassware, use a ceramic-centered wire gauze to protect glassware from direct contact with the flame. Wire gauzes can withstand extremely high temperatures and will help prevent glassware from breaking. **Figure D** shows the proper setup for evaporating a solution over a water bath.

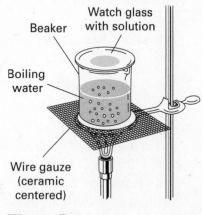

Watch glass with solution

Beaker

Boiling water

Wire gauze (ceramic centered)

Figure D

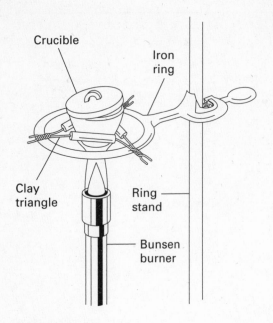

Figure E

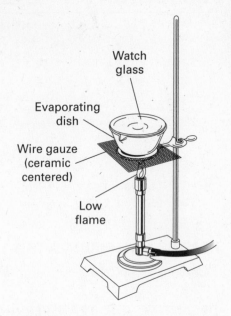

Figure F

3. In some experiments you are required to heat a substance to high temperatures in a porcelain crucible. **Figure E** shows the proper apparatus setup used to accomplish this task.

4. **Figure F** shows the proper setup for evaporating a solution in a porcelain evaporating dish with a watch glass cover that prevents spattering.

5. Glassware, porcelain, and iron rings that have been heated may look cool after they are removed from a heat source, but they can burn your skin even after several minutes of cooling. Use tongs, test tube holders, or heat-resistant mitts and pads whenever you handle this apparatus.

6. You can test the temperature of questionable beakers, ring stands, wire gauzes, or other pieces of apparatus that have been heated, by holding the back of your hand close to their surfaces before grasping them. You will be able to feel any heat generated from the hot surfaces. **Do not touch the apparatus until it is cool.**

POURING LIQUID FROM A REAGENT BOTTLE

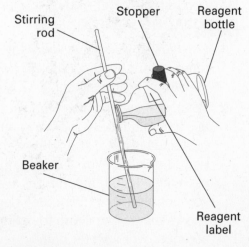

Figure G

1. Read the label at least three times before using the contents of a reagent bottle.

2. Never lay the stopper of a reagent bottle on the lab table.

3. When pouring a caustic or corrosive liquid into a beaker, use a stirring rod to avoid drips and spills. Hold the stirring rod against the lip of the reagent bottle. Estimate the amount of liquid you need, and pour this amount along the rod into the beaker. See **Figure G.**

4. Take extra precautions when handling a bottle of acid or strong base. Remember the following important rules: Never add water to any concentrated acid, particularly sulfuric acid, because the mixture can splash and will generate a lot of heat. To dilute any acid, add the acid to water in small quantities, while stirring slowly. Remember the "triple A's"—Always Add Acid to water.

5. Examine the outside of the reagent bottle for any liquid that has dripped down the bottle or spilled on the counter top. Your teacher will show you the proper procedures for cleaning up a chemical spill.

6. Never pour reagents back into stock bottles. At the end of the experiment, your teacher will tell you how to dispose of any excess chemicals.

HEATING MATERIAL IN A TEST TUBE

1. Check to see that the test tube is heat resistant.

2. Always use a test-tube holder or clamp when heating a test tube.

3. Never point a heated test tube at anyone, because the liquid may splash out of the test tube.

4. Never look down into the test tube while heating it.

5. Heat the test tube from the upper portions of the tube downward and continuously move the test tube, as indicated in **Figure H.** Do not heat any one spot on the test tube. Otherwise, a pressure buildup may cause the bottom of the tube to blow out.

USING A MORTAR AND PESTLE

1. A mortar and pestle should be used for grinding only one substance at a time. See **Figure I.**

2. Never use a mortar and pestle for simultaneously mixing different substances.

3. Place the substance to be broken up into the mortar.

4. Firmly push on the pestle to crush the substance. Then grind it to pulverize it.

5. Remove the powdered substance with a porcelain spoon.

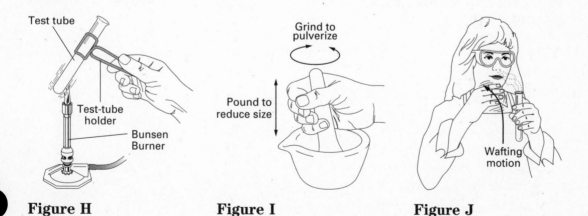

Figure H   **Figure I**   **Figure J**

DETECTING ODORS SAFELY

1. Test for the odor of gases by wafting your hand over the test tube and cautiously sniffing the fumes, as indicated in **Figure J.**

2. Do not inhale any fumes directly.

3. Use a fume hood whenever poisonous or irritating fumes are involved. **Do not** waft and sniff poisonous or irritating fumes.

Inquiry

Mixture Separation

The ability to separate and recover pure substances from mixtures is extremely important in scientific research and industry. Chemists need to work with pure substances, but naturally occurring materials are seldom pure. Often, differences in the physical properties of the components in a mixture provide the means for separating them. In this experiment, you will have an opportunity to design, develop, and implement your own procedure for separating a mixture. The mixture you will work with contains salt, sand, iron filings, and poppy seeds. All four substances are in dry, granular form.

OBJECTIVES

• **Observe** the chemical and physical properties of a mixture.

• **Relate** knowledge of chemical and physical properties to the task of purifying the mixture.

• **Analyze** the success of methods of purifying the mixture.

MATERIALS

• aluminum foil
• cotton balls
• distilled water
• filter funnels
• filter paper
• forceps
• magnet
• paper clips
• paper towels
• Petri dish
• pipets

• plastic forks
• plastic spoons
• plastic straws
• rubber stoppers
• sample of mixture and components (sand, iron filings, salt, poppy seeds)
• test tubes and rack
• tissue paper
• transparent tape
• wooden splints

Always wear safety goggles and a lab apron to protect your eyes and clothing. If you get a chemical in your eyes, immediately flush the chemical out at the eyewash station while calling to your teacher. Know the locations of the emergency lab shower and the eyewash station and the procedures for using them.

PREPARATION

Your task will be to plan and carry out the separation of a mixture. Before you can plan your experiment, you will need to investigate the properties of each component in the mixture. The properties will be used to design your mixture separation.

Mixture Separation *continued*

PROCEDURE

1. Obtain separate samples of each of the four mixture components from your teacher. Use the equipment you have available to make observations of the components and determine their properties. You will need to run several tests with each substance, so don't use all of your sample on the first test. Look for things like whether the substance is magnetic, whether it dissolves, or whether it floats. Record your observations in the **Data Table.**

2. Make a plan for what you will do to separate a mixture that includes the four components from step **1.** Review your plan with your teacher.

3. Obtain a sample of the mixture from your teacher. Using the equipment you have available, run the procedure you have developed.

DISPOSAL

4. Clean your lab station. Clean all equipment, and return it to its proper place. Dispose of chemicals and solutions in the containers designated by your teacher. Do not pour any chemicals down the drain or throw anything in the trash unless your teacher directs you to do so. Wash your hands thoroughly after all work is finished and before you leave the lab.

Data Table				
Properties	**Sand**	**Iron filings**	**Salt**	**Poppy seeds**
Dissolves	**no**	**no**	**yes**	**no**
Floats	**no**	**no**	**no**	**yes**
Magnetic	**no**	**yes**	**no**	**no**
Other	**sharp**	**sharp**	**white**	**round**

Analysis

1. **Evaluating Methods** On a scale of 1 to 10, how successful were you in separating and recovering each of the four components: sand, salt, iron filings, and poppy seeds? Consider 1 to be the best and 10 to be the worst. Justify your ratings based on your observations.

   Students' answers will vary. They may justify their estimations of success by

   the presence (or lack) of impurities in the separated components and by the

   amount recovered.

   _____

   _____

Conclusions

1. **Evaluating Methods** How did you decide on the order of your procedural steps? Would any order have worked?

   Students' answers will vary.

2. **Designing Experiments** If you could do the lab over again, what would you do differently? Be specific.

   Students' answers will vary but should be logical and demonstrate students'

   understanding of the physical properties of each substance.

3. **Designing Experiments** Name two materials or tools that weren't available that might have made your separation easier.

   Students' answers will vary. Note the suggestions students make and

   consider making those items available the next time you use this experiment.

4. **Applying Ideas:** For each of the four components, describe a specific physical property that enabled you to separate the component from the rest of the mixture.

   Students' answers will vary but may include magnetism, density, and solubility:

   iron is attracted by a magnet, salt dissolves in water, poppy seeds float on

   water, and sand sinks in water.

Name _____ Class _____ Date _____

Mixture Separation *continued*

EXTENSIONS

1. Evaluating Methods What methods could be used to determine the purity of each of your recovered components?

Students' answers will vary but may include comparing the densities of their

samples with the densities of standard samples of the pure components and

picking through small samples from within their samples to check for visible

impurities.

_____

_____

2. Designing Experiments How could you separate each of the following two-part mixtures? **Students' answers will vary but may include the following:**

a. aluminum filings and iron filings

using a magnet

b. sand and gravel

pouring the mixture through a screen or sifter

c. sand and finely ground polystyrene foam

adding water and checking for flotation

d. salt and sugar

dissolving the mixture in a solvent that dissolves one component but

not the other

e. alcohol and water

distillation

f. nitrogen and oxygen

distillation

3. Designing Experiments One of the components of the mixture in this experiment is in a different physical state at the completion of this experiment than it was at the start. Which one? How would you convert that component back to its original state?

___The table salt, which was a solid at the start of the experiment, is in solution___

___in the water at the end of the experiment. The salt could be restored to its___

___original solid state by evaporating the water from the solution.___

_____

_____

Quick Lab
Density of Pennies

MATERIALS

• balance

• 100 mL graduated cylinder

• 40 pennies dated before 1982

• 40 pennies dated after 1982

• water

 Always wear safety goggles and a lab apron to protect your eyes and clothing. If you get a chemical in your eyes, immediately flush the chemical out at the eyewash station while calling to your teacher. Know the locations of the emergency lab shower and the eyewash station and the procedures for using them.

PROCEDURE

1. Using the balance, determine the mass of the 40 pennies minted prior to 1982. Repeat this measurement two more times. Average the results of the three trials to determine the average mass of the pennies.

2. Repeat step **1** with the 40 pennies minted after 1982.

3. Pour about 50 mL of water into the 100 mL graduated cylinder. Record the exact volume of the water. Add the 40 pennies minted before 1982. CAUTION: Add the pennies carefully so that no water is splashed out of the cylinder. Record the exact volume of the water and pennies. Repeat this process two more times. Determine the volume of the pennies for each trial. Average the results of those trials to determine the average volume of the pennies.

4. Repeat step **3** with the 40 pennies minted after 1982.

5. Review your data for any large differences between trials that could increase the error of your results. Repeat those measurements.

6. Use the average volume and average mass to calculate the average density for each group of pennies.

7. Compare the calculated average densities with the density of the copper listed in **Table 4** on page 38 of the textbook.

DISCUSSION

1. Why is it best to use the results of three trials rather than a single trial for determining the density?

   <u>The amount by which the result will be affected by a measurement error is</u>

   <u>decreased.</u>

   _____

   _____

Density of Pennies *continued*

2. How did the densities of the two groups of pennies compare? How do you account for any difference?

The post-1982 pennies have a lower density. The composition of the pre- and

post-1982 pennies differs, resulting in differing masses.

_____

3. Use the results of this investigation to formulate a hypothesis about the composition of the two groups of pennies. How could you test your hypothesis?

Hypotheses will vary. Students should know that pennies contain copper.

Research should show that they also contain zinc. Students could find the

densities of copper and zinc and conclude that the less dense pennies

contain more zinc and less copper because zinc is less dense than copper.

_____

Name _____ Class _____ Date _____

Percentage of Water in Popcorn

Popcorn pops because of the natural moisture inside each kernel. When the internal water is heated above 100°C, the liquid water changes to a gas, which takes up much more space than the liquid, so the kernel expands rapidly.

The percentage of water in popcorn can be determined by the following equation.

$$\frac{initial\ mass - final\ mass}{initial\ mass} \times 100 = percentage\ of\ H_2O\ in\ unpopped\ popcorn$$

The popping process works best when the kernels are first coated with a small amount of vegetable oil. Make sure you account for the presence of this oil when measuring masses. In this lab, you will design a procedure for determining the percentage of water in three samples of popcorn. The popcorn is for testing only, and *must not* be eaten.

OBJECTIVES

- **Measure** the masses of various combinations of a beaker, oil, and popcorn kernels.

- **Determine** the percentages of water in popcorn kernels.

- **Compare** experimental data.

MATERIALS

- aluminum foil (1 sheet)

- beaker, 250 mL

- Bunsen burner with gas tubing and striker

- kernels of popcorn for each of three brands (80)

- oil to coat the bottom of the beaker

- ring stand, iron ring, and wire gauze

Always wear safety goggles and a lab apron to protect your eyes and clothing. If you get a chemical in your eyes, immediately flush the chemical out at the eyewash station while calling to your teacher. Know the locations of the emergency lab shower and the eyewash station and the procedures for using them.

When using a Bunsen burner, confine long hair and loose clothing. If your clothing catches on fire, WALK to the emergency lab shower and use it to put out the fire. When heating a substance in a test tube, the mouth of the test tube should point away from where you and others are standing. Watch the test tube at all times to prevent the contents from boiling over.

Name _____ Class _____ Date _____

Percentage of Water in Popcorn continued

PREPARATION

Use the **Data Table** provided to record your data.

PROCEDURE

1. Measure the mass of a 250 mL beaker. Record the mass in the **Data Table.**

2. Add a small amount of vegetable oil to the beaker to coat the bottom of it. Measure the mass of the beaker and oil. Record the mass in the **Data Table.**

3. Add 20 kernels of brand A popcorn to the beaker. Shake the beaker gently to coat the kernels with oil. Measure the mass of the beaker, oil, and popcorn. Record the mass in the **Data Table.**

4. Subtract the mass found in step **2** from the mass found in step **3** to obtain the mass of 20 unpopped kernels. Record the mass in the **Data Table.**

5. Cover the beaker loosely with the aluminum foil. Punch a few small holes in the aluminum foil to let moisture escape. These holes should not be large enough to let the popping corn pass through.

6. Heat the popcorn until the majority of the kernels have popped. The popcorn pops more efficiently if the beaker is held firmly with tongs and gently shaken side to side on the wire gauze.

7. Remove the aluminum foil from the beaker and allow the beaker to cool for 10 minutes. Then, measure the mass of the beaker, oil, and popped corn. Record the mass in the **Data Table.**

8. Subtract the mass in step **7** from the mass in step **3** to obtain the mass of water that escaped when the corn popped. Record the mass in the **Data Table.**

9. Calculate the percentage of water in the popcorn.

10. Dispose of the popcorn in the designated container. Remove the aluminum foil, and set it aside. Clean the beaker, and dry it well. Alternatively, if your teacher approves, use a different 250 mL beaker.

11. Repeat steps **1–10** for brand B popcorn.

12. Repeat steps **1–10** for brand C popcorn.

DISPOSAL

13. Dispose of popped popcorn and aluminum foil in containers as directed by your instructor. Do not eat the popcorn.

14. Clean beakers. Return beakers and other equipment to the proper place.

15. Clean all work surfaces and personal protective equipment as directed by your instructor.

16. Wash your hands thoroughly before leaving the laboratory.

Name _____ Class _____ Date _____

Percentage of Water in Popcorn *continued*

Data Table			
	Popcorn Brand A	Popcorn Brand B	Popcorn Brand C
Mass of 250 mL beaker (g)	100.27 g	112.44 g	98.18 g
Mass of beaker + oil (g)	102.74 g	114.35 g	99.73 g
Mass of beaker + oil + 20 kernels (before) (g)	104.77 g	117.82 g	103.22 g
Mass of 20 kernels before (g)	2.03 g	3.47 g	3.49 g
Mass of beaker + oil + 20 kernels (after) (g)	104.33 g	117.32 g	102.78 g
Mass of 20 kernels (after) (g)	1.59 g	2.97 g	3.05 g
Mass of water in 20 kernels (g)	0.44 g	0.50 g	0.44 g
Percentage of water in popcorn	22%	14%	13%

Analysis

1. **Applying Ideas:** Determine the mass of the 20 unpopped kernels of popcorn for each brand of popcorn.

   Student answers will depend on the sample.

   _____

   _____

2. **Applying Ideas:** Determine the mass of the 20 popped kernels of popcorn for each brand of popcorn.

   Student answers will depend on the sample.

   _____

   _____

3. **Applying Ideas:** Determine the mass of the water that was lost when the popcorn popped for each brand.

   Student answers will depend on the sample.

   _____

   _____

Name _____ Class _____ Date _____

Percentage of Water in Popcorn *continued*

Conclusions

1. Applying Data: Determine the percentage by mass of water in each brand of popcorn.

Student answers will depend on the sample. Typical answers will probably

range up to 25% but could be substantially less if the popcorn has been

stored for a long period of time.

_____

_____

2. Inferring Relationships: Do all brands of popcorn contain the same percentage water?

Students should conclude that the percentage of water varies between

brands of popcorn.

_____

_____

EXTENSIONS

1. Designing Experiments: What are some likely areas of imprecision in this experiment?

Student answers will vary. One possibility is failure to dry the beaker

between experiments, which may cause some of the residual water not in the

popcorn sample to be lost; this would give an answer that is too high. If the

sample is not heated sufficiently, some kernels may not pop, which would

give an answer that is too low. A common mistake might be to forget to

remove the aluminum foil before the final mass measurement; this could give

an answer that is too low or even negative (physically impossible).

_____

_____

Percentage of Water in Popcorn *continued*

2. Designing Experiments: Do you think that the volume of popped corn depends on the percentage of water in the unpopped corn? Design an experiment to find the answer.

Student answers will vary. Student procedures should be evaluated according

to organization and accuracy.

_____

_____

_____

_____

_____

_____

_____

_____

_____

Name _____ Class _____ Date _____

Quick Lab

DATASHEET FOR IN-TEXT LAB

Constructing a Model

How can you construct a model of an unknown object by (1) making inferences about an object that is in a closed container and (2) touching the object without seeing it?

MATERIALS

- can covered by a sock sealed with tape
- one or more objects that fit in the container
- metric ruler
- balance

Always wear safety goggles and a lab apron to protect your eyes and clothing. If you get a chemical in your eyes, immediately flush the chemical out at the eyewash station while calling to your teacher. Know the locations of the emergency lab shower and the eyewash station and the procedures for using them.

PROCEDURE

Record all of your results in the **Data Table.**

1. Your teacher will provide you with a can that is covered by a sock sealed with tape. Without unsealing the container, try to determine the number of objects inside the can as well as the mass, shape, size, composition, and texture of each. To do this, you may carefully tilt or shake the can. Record your observations in the **Data Table.**

2. Remove the tape from the top of the sock. Do not look inside the can. Put one hand through the opening, and make the same observations as in step **1** by handling the objects. To make more-accurate estimations, practice estimating the sizes and masses of some known objects outside the can. Then compare your estimates of these objects with actual measurements using a metric ruler and a balance.

Data Table	
Observations	
Sealed can	**Unsealed can**
Observations will vary.	Observations will vary.

DISCUSSION

1. Scientists often use more than one method to gather data. How was this illustrated in the investigation?

   Scientists use both indirect and direct observations to gather data.

   In part 1, observations were made without contact with the objects themselves.

   These observations were indirect. In part 2, students made direct observations

   by touching the objects. Between the two types of observations, students could

   get a good idea of the identity of the objects.

   _____

2. Of the observations you made, which were qualitative and which were quantitative?

   Determinations of the mass, size, and number of objects are quantitative.

   Shape, material, and texture determinations are qualitative.

   _____

   _____

3. Using the data you gathered, draw a model of the unknown object(s) and write a brief summary of your conclusions.

   Answers will vary.

   _____

   _____

   _____

   _____

Conservation of Mass

The law of conservation of mass states that matter is neither created nor destroyed during a chemical reaction. Therefore, the mass of a system should remain constant during any chemical process. In this experiment, you will determine whether mass is conserved by examining a simple chemical reaction and comparing the mass of the system before the reaction with its mass after the reaction.

OBJECTIVES

- **Observe** the signs of a chemical reaction.
- **Compare** masses of reactants and products.
- **Design** experiments.
- **Relate** observations to the law of conservation of mass.

MATERIALS

- 2 L plastic soda bottle
- 5% acetic acid solution (vinegar)
- balance
- clear plastic cups, 2
- graduated cylinder
- hook-insert cap for bottle
- microplunger
- sodium hydrogen carbonate (baking soda)

Always wear safety goggles and a lab apron to protect your eyes and clothing. If you get a chemical in your eyes, immediately flush the chemical out at the eyewash station while calling to your teacher. Know the locations of the emergency lab shower and the eyewash station and the procedures for using them.

PREPARATION

Use the data tables provided to record your data and observations for Part I and Part II.

PROCEDURE—PART I

1. Obtain a microplunger, and tap it down into a sample of baking soda until the bulb end is packed with a plug of the powder (4–5 mL of baking soda should be enough to pack the bulb).

2. Hold the microplunger over a plastic cup, and squeeze the sides of the microplunger to loosen the plug of baking soda so that it falls into the cup.

3. Use a graduated cylinder to measure 100 mL of vinegar, and pour it into a second plastic cup.

Conservation of Mass *continued*

4. Place the two cups side by side on the balance pan, and measure the total mass of the system (before reaction) to the nearest 0.01 g. Record the mass in **Data Table-Part I.**

5. Add the vinegar to the baking soda a little at a time to prevent the reaction from getting out of control. Allow the vinegar to slowly run down the inside of the cup. Observe and record your observations about the reaction.

6. When the reaction is complete, place both cups on the balance, and determine the total final mass of the system to the nearest 0.01 g. Calculate any change in mass. Record both the final mass and any change in mass in **Data Table-Part I.**

7. Examine the plastic bottle and the hook-insert cap. Try to develop a modified procedure that will test the law of conservation of mass more accurately than the procedure in Part I.

8. Write the answers to items 1 through 3 in Analysis—Part I.

PROCEDURE—PART II

9. Your teacher should approve the procedure you designed in Procedure—Part I, step **7.** Implement your procedure with the same chemicals and quantities you used in Part I, but use the bottle and hook-insert cap in place of the two cups. Record your data in **Data Table-Part II.**

10. If you were successful in step **9** and your results reflect the conservation of mass, proceed to complete the experiment. If not, find a lab group that was successful, and discuss with them what they did and why they did it. Your group should then test the other group's procedure to determine whether their results are reproducible.

DISPOSAL

11. Clean your lab station. Clean all equipment, and return it to its proper place. Dispose of chemicals and solutions in the containers designated by your teacher. Do not pour any chemicals down the drain or throw anything in the trash unless your teacher directs you to do so. Wash your hands thoroughly after all work is finished and before you leave the lab.

❚ Conservation of Mass *continued*

Data Table-Part I

Initial mass (g)	Final mass (g)	Change in mass (g)	Observations
*	*	*	solution fizzes; baking soda disappears

Data Table-Part II

Initial mass (g)	Final mass (g)	Change in mass (g)	Observations
*	*	*	solution bubbles; fizzing noise not as loud; sides of bottle tight; hissing noise when cap removed

* Student data will vary.

Analysis
PART I

1. Drawing Conclusions What evidence was there that a chemical reaction occurred?

The reactants effervesce with the formation of a new gaseous substance. The

baking soda seemed to disappear eventually.

2. Organizing Data How did the final mass of the system compare with the initial mass of the system?

Students' answers will vary, but typical results show a mass loss of 1.0–1.5 g

for Part I.

| Conservation of Mass *continued*

3. **Resolving Discrepancies** Does your answer to the previous question show that the law of conservation of mass was violated? (Hint: Another way to express the law of conservation of mass is to say that the mass of all of the products equals the mass of all of the reactants.) What do you think might cause the mass difference?

Students' answers should indicate that although the mass loss appears

to violate the law of conservation of mass, the law probably still holds.

The reason is that not all of the products were present for the second

measurement of mass. The bubbles that formed and popped were a gaseous

product that escaped into the air.

_____

Analysis
PART II

1. **Drawing Conclusions** Was there any new evidence in Part II indicating that a chemical reaction occurred?

Again, there was vigorous bubbling for the first few seconds. However, the

fizzing noise was not very audible. The sides of the bottle became tight and

hard to push in, and when the cap was removed, the hissing noise of gas

escaping under pressure could be heard.

_____

_____

2. **Organizing Ideas** Identify the state of matter for each reactant in Part II. Identify the state of matter for each product.

The reactants were a liquid (vinegar) and a solid (baking soda). The

products of the reaction were a gas (the carbon dioxide seen forming the

bubbles) and a liquid (a solution of excess vinegar and sodium acetate).

_____

_____

Conservation of Mass *continued*

Conclusions

1. Relating Ideas What is the difference between the system in Part I and the system in Part II? What change led to the improved results in Part II?

In Part I the system was left open, but it was closed in Part II so that not

even gas could escape.

_____

2. Evaluating Methods Why did the procedure for Part II work better than the procedure for Part I?

Part II worked better because one of the products was a gas and the bottle-

and-cap system kept the gas trapped so that its mass could be measured with

the other products. In Part I, the gas escaped.

_____

EXTENSIONS

1. Applying Models When a log burns, the resulting ash obviously has less mass than the unburned log did. Explain whether this loss of mass violates the law of conservation of mass.

Just as the reaction studied in this experiment produced a product that

escaped into the environment, the burning of a log produces many products,

not just ashes. The smoke, CO_2, and water vapor escape as the log burns.

Presumably, if one had a way to measure the total mass of the products, it

would be the same as the total mass of the reactants.

_____

_____

_____

Conservation of Mass *continued*

2. **Designing Experiments** Design a procedure that would test the law of conservation of mass for the burning log described in Extension item 1.

Students' answers will vary, but they should describe a way to burn a log in a

closed, rigid container that could withstand heat and flames without reacting.

Students may also recognize that the amount of oxygen needed to burn the

log would require a very large container unless the oxygen were purified from

the air or kept under high pressure. You may want to point out to students

that this experiment would be very difficult and potentially dangerous

because of the pressures that would build up as a result of the carbon in the

log being converted to carbon dioxide gas.

_____

_____

_____

Name _____ Class _____ Date _____

The Wave Nature of Light: Interference

Does light show the wave property of interference when a beam of light is projected through a pinhole onto a screen?

MATERIALS

- scissors
- manila folders
- thumbtack
- masking tape

- aluminum foil
- white poster board or cardboard
- flashlight

PROCEDURE

Record all your observations.

1. To make the pinhole screen, cut a 20 cm × 20 cm square from a manila folder. In the center of the square, cut a 2 cm square hole. Cut a 7 cm × 7 cm square of aluminum foil. Using a thumbtack, make a pinhole in the center of the foil square. Tape the aluminum foil over the 2 cm square hole, making sure the pinhole is centered.

2. Use white poster board to make a projection screen 35 cm × 35 cm.

3. In a dark room, center the light beam from a flashlight on the pinhole. Hold the flashlight about 1 cm from the pinhole. The pinhole screen should be about 50 cm from the projection screen. Adjust the distance to form a sharp image on the projection screen.

DISCUSSION

1. Did you observe interference patterns on the screen?

   Most students should observe light and dark rings around the edge of the

   hole illuminated on the screen. These light and dark patterns are a result

   of interference.

   _____

2. As a result of your observations, what do you conclude about the nature of light?

   Light has wavelike properties.

   _____

   _____

Flame Tests

The characteristic light emitted by an element is the basis for the chemical test known as a *flame test*.

To identify an unknown substance, you must first determine the characteristic colors produced by different elements. You will do this by performing a flame test on a variety of standard solutions of metal compounds. Then, you will perform a flame test with an unknown sample to see if it matches any of the standard solutions. The presence of even a speck of another substance can interfere with the identification of the true color of a particular type of atom, so be sure to keep your equipment very clean and perform multiple trials to check your work.

OBJECTIVES

- **Identify** a set of flame-test color standards for selected metal ions.

- **Relate** the colors of a flame test to the behavior of excited electrons in a metal ion.

- **Identify** an unknown metal ion by using a flame test.

- **Demonstrate** proficiency in performing a flame test and in using a spectroscope.

MATERIALS

- 250 mL beaker
- Bunsen burner and related equipment
- cobalt glass plates
- crucible tongs
- distilled water
- flame-test wire
- glass test plate (or a microchemistry plate with wells)
- spectroscope

- 1.0 M HCl solution
- $CaCl_2$ solution
- K_2SO_4 solution
- Li_2SO_4 solution
- Na_2SO_4 solution
- $SrCl_2$ solution
- unknown solution
- wooden splints (optional)

Always wear safety goggles and a lab apron to protect your eyes and clothing. If you get a chemical in your eyes, immediately flush the chemical out at the eyewash station while calling to your teacher. Know the locations of the emergency lab shower and the eyewash station and the procedures for using them.

Flame Tests *continued*

⚠ **Do not touch any chemicals.** If you get a chemical on your skin or clothing, wash the chemical off at the sink while calling to your teacher. Make sure you carefully read the labels and follow the precautions on all containers of chemicals that you use. If there are no precautions stated on the label, ask your teacher what precautions you should follow. Do not taste any chemicals or items used in the laboratory. Never return leftovers to their original container; take only small amounts to avoid wasting supplies.

⚠ **When using a Bunsen burner, confine long hair and loose clothing.** If your clothing catches on fire, WALK to the emergency lab shower and use it to put out the fire. When heating a substance in a test tube, the mouth of the test tube should point away from where you and others are standing. Watch the test tube at all times to prevent the contents from boiling over.

PREPARATION

1. Use the **Data Table** provided to record your data.

2. Label a beaker *Waste*. Thoroughly clean and dry a well strip. Fill the first well one-fourth full with 1.0 M HCl on the plate. Clean the test wire by first dipping it in the HCl and then holding it in the colorless flame of the Bunsen burner. Repeat this procedure until the flame is not colored by the wire. When the wire is ready, rinse the well with distilled water and collect the rinse water in the waste beaker.

3. Put 10 drops of each metal ion solution listed in the materials list in a row in each well of the well strip. Put a row of 1.0 M HCl drops on a glass plate across from the metal ion solutions. Record the positions of all of the chemicals placed in the wells. The wire will need to be cleaned thoroughly between each test solution with HCl to avoid contamination from the previous test.

PROCEDURE

1. Dip the wire into the $CaCl_2$ solution, and then hold it in the Bunsen burner flame. Observe the color of the flame, and record it in the **Data Table.** Repeat the procedure again, but this time look through the spectroscope to view the results. Record the wavelengths you see from the flame. Repeat each test three times. Clean the wire with the HCl as you did in Preparation step **2.**

2. Repeat step **1** with the K_2SO_4 and with each of the remaining solutions in the well strip.

3. Test another drop of Na_2SO_4, but this time view the flame through two pieces of cobalt glass. Clean the wire, and repeat the test. Record in the **Data Table** the colors and wavelengths of the flames as they appear when viewed through the cobalt glass. Clean the wire and the well strip, and rinse the well strip with distilled water. Pour the rinse water into the waste beaker.

4. Put a drop of K_2SO_4 in a clean well. Add a drop of Na_2SO_4. Perform a flame test for the mixture. Observe the flame without the cobalt glass. Repeat the test again, but this time observe the flame through the cobalt glass. Record in the **Data Table** the colors and wavelengths of the flames. Clean the wire, and rinse the well strip with distilled water. Pour the rinse water into the waste beaker.

5. Obtain a sample of the unknown solution. Perform flame tests for it with and without the cobalt glass. Record your observations. Clean the wire, and rinse the well strip with distilled water. Pour the rinse water into the waste beaker.

DISPOSAL

6. Dispose of the contents of the waste beaker in the container designated by your teacher. Wash your hands thoroughly after cleaning up the area and equipment.

Data Table		
Metal compound	**Color of flame**	**Wavelengths detected (nm)**
$CaCl_2$	yellowish red (orange)	420, 445, 460, 485, 610, 645, 650
K_2SO_4	violet (purple)	405, 408, 695, 700
Li_2SO_4	red (carmine)	462, 498, 612, 670
Na_2SO_4	yellow	590, 595
$SrCl_2$	scarlet	405, 420, 460, 485, 490, 500, 665, 685, 710
Na only (cobalt glass)	only blue of the glass is visible	
K only (cobalt glass)	violet (purple)	
Na and K	yellow	
Na and K (cobalt glass)	violet	
Unknown	**Answers will vary.**	

Flame Tests *continued*

Analysis

1. **Organizing Data** Examine your data table, and create a summary of the flame test for each metal ion.

   Note: Assign only Analysis items 1–3 if spectroscopes are unavailable.

   See sample data table.

   _____

   _____

2. **Analyzing Data** Account for any differences in the individual trials for the flame tests for the metals ions.

   Student answers will vary. Some students may have had difficulty properly

   cleaning the wire, so the first test of a new compound may have traces of the

   previous one.

   _____

   _____

3. **Organizing Ideas** Explain how viewing the flame through cobalt glass can make it easier to analyze the ions being tested.

   The flame color of potassium is purple, but it is so weak that it can be

   overpowered by the yellow sodium light if a mixture is tested. The cobalt

   glass screens out the yellow sodium light.

   _____

   _____

4. **Relating Ideas** For three of the metal ions tested, explain how the flame color you saw relates to the lines of color you saw when you looked through the spectroscope.

   Answers will vary, but students should realize that the colors seen by the eye

   were the result of combining the colors of light seen in the line spectra.

   _____

   _____

   _____

Conclusions

1. **Inferring Conclusions** What metal ions are in the unknown solution?

   Answers will vary. Students should be able to identify the unknown by

   comparing its results with the results for the other metal compounds tested.

2. **Evaluating Methods** How would you characterize the flame test with respect to its sensitivity? What difficulties could there be when identifying ions by the flame test?

   The flame test is fairly specific because it can show an easily detectable

   signal with a very small amount of material. Possible difficulties include

   problems with contamination and the fact that some metals have similar

   colors when flame tested.

EXTENSIONS

1. **Inferring Conclusions** A student performed flame tests on several unknowns and observed that they all were shades of red. What should the student do to correctly identify these substances? Explain your answer.

   The student should compare the shades of red with the colors of the known

   samples. If information about spectral lines is available, it would also help

   determine which metal is the unknown.

2. **Applying Ideas** During a flood, the labels from three bottles of chemicals were lost. The three unlabeled bottles of white solids were known to contain the following: strontium nitrate, ammonium carbonate, and potassium sulfate. Explain how you could easily test the substances and relabel the three bottles. (Hint: Ammonium ions do not provide a distinctive flame color.)

   Strontium nitrate will change the color of the flame to red, potassium sul-

   fate will change the flame color to purple, and ammonium will not change the

   flame color.

Designing Your Own Periodic Table

Can you design your own periodic table using information similar to that available to Mendeleev?

MATERIALS

- index cards

PROCEDURE

1. Write down the information available for each element on separate index cards. The following information is appropriate: a letter of the alphabet (A, B, C, etc.) to identify each element; atomic mass; state; density; melting point; boiling point; and any other readily observable physical properties. Do not write the name of the element on the index card, but keep a separate list indicating the letters you have assigned to each element.

2. Organize the cards for the elements in a logical pattern as you think Mendeleev might have done.

DISCUSSION

1. Keeping in mind that the information you have is similar to that available to Mendeleev in 1869, answer the following questions.

   a. Why are atomic masses given instead of atomic numbers?

   Atomic numbers were not discovered until about 45 years later.

   _____

   b. Can you identify each element by name?

   Answers may vary.

   _____

2. How many groups of elements, or families, are in your periodic table? How many periods, or series, are in the table?

   Answers may vary.

   _____

3. Predict the characteristics of any missing elements. When you have finished, check your work using your separate list of elements and a periodic table.

   Answers may vary.

   _____

Name _____ Class _____ Date _____

The Mendeleev Lab of 1869

Russian chemist Dmitri Mendeleev is generally credited with being the first chemist to observe that patterns emerge when the elements are arranged according to their properties. Mendeleev's arrangement of the elements was unique because he left blank spaces for elements that he claimed were undiscovered as of 1869. Mendeleev was so confident that he even predicted the properties of these undiscovered elements. His predictions were eventually proven to be quite accurate, and these new elements fill the spaces that originally were blank in his table. Use your knowledge of the periodic table to determine the identity of each of the nine unknown elements in this activity. These unknown elements are from the periodic table's groups that are listed below. Each of these groups contains at least one unknown element.

1   2   11   13   14   17   18

None of the known elements serves as one of the nine unknown elements. No radioactive elements are used during this experiment. The relevant radioactive elements include Fr, Ra, At, and Rn. You may not use your textbook or other reference materials. You have been provided with enough information to determine each of the unknown elements.

OBJECTIVES

- **Observe** the physical properties of common elements.

- **Observe** the properties and trends in the elements on the periodic table.

- **Draw conclusions** and **identify** unknown elements based on observed trends in properties.

MATERIALS

- blank periodic table

- elemental samples: Ar, C, Sn, and Pb

- note cards, 3 in. × 5 in.

- periodic table with element names and symbols, period numbers, and group numbers

- element cards

Always wear safety goggles and a lab apron to protect your eyes and clothing. If you get a chemical in your eyes, immediately flush the chemical out at the eyewash station while calling to your teacher. Know the locations of the emergency lab shower and the eyewash station and the procedures for using them.

The Mendeleev Lab of 1869 *continued*

PREPARATION

1. Use the **Data Table** provided to record the properties of each unknown that you test.

2. Use the note cards to copy the information listed on each of the sample cards. If the word *observe* is listed, you will need to visually inspect the sample and then write the observation in the appropriate space.

PROCEDURE

1. Arrange the note cards of the known elements in a rough representation of the periodic table. In other words, all of the known elements from Group 1 should be arranged in the appropriate order. Arrange all of the other cards accordingly.

2. Inspect the properties of the unknowns to see where properties would best "fit" the trends of the elements of each group.

3. Assign the proper element name to each of the unknowns. Add the symbol for each one of the unknown elements to your data table.

DISPOSAL

4. Clean up your lab station, and return the leftover note cards and samples of the elements to your teacher. Do not pour any of the samples down the drain or place them in the trash unless your teacher directs you to do so. Wash your hands thoroughly before you leave the lab and after all your work is finished.

Data Table		
Unknown	**Properties**	**Element**
Number of unknowns will vary.	Observations will vary.	Answers will vary.

Analysis

1. **Organizing Ideas** In what order did your group arrange the properties to determine the unknowns? Explain your reasoning. Would a different order have been better? If so, what is the better order and why?

   <u>Answers may vary. One possible answer: First, use physical state because all</u>

   <u>gases are nonmetals, although not all solids are metals. Second, finish</u>

   <u>determining the gases by looking at the color. Then, examine the solids by</u>

   <u>comparing the conductivity; the greatest conductivity difference will reflect</u>

   <u>differences between metals versus semi-metals or nonmetals. Last, compare</u>

   <u>their reactivity in water, color, and melting point.</u>

   _____

   _____

   _____

2. **Evaluating Methods** What properties were the most useful in sorting the unknowns? What properties were the least useful? Explain your answer.

   <u>Answers may vary. The most useful properties should be physical state, color,</u>

   <u>and conductivity. These properties quickly separated the unknowns into</u>

   <u>metals, semi-metals, and nonmetals. These properties also helped distinguish</u>

   <u>between the metals. The least useful properties were density and hardness.</u>

   <u>The unknowns could be determined without these two properties. These</u>

   <u>properties were also difficult to determine a trend until after the unknown</u>

   <u>was placed in its position in the table.</u>

   _____

   _____

   _____

| The Mendeleev Lab of 1869 *continued*

Conclusions

1. **Interpreting Information** Summarize your group's reasoning for the assignment of each unknown. Explain in a few sentences exactly how you predicted the identity of the nine unknown elements.

   Answers may vary. _____

   _____

   _____

   _____

   _____

   _____

   _____

EXTENSIONS

1. **Predicting Outcomes** Use only the data from your group's experiment to predict the properties of the not yet discovered element, which has an atomic number of 120 (assuming it does not radioactively decay).

   The element with atomic number 120 would be found in Group 2, which

   consists of the alkaline-earth metals. The trends seen from the data in this

   experiment suggest that some of its properties might be the following: reacts

   very strongly with water, is a solid, conducts well, and is some shade of

   silvery white. _____

   _____

   _____

   _____

Types of Bonding in Solids

The purpose of this experiment is to relate certain properties of solids to the type of bonding the solids have. These observable properties depend on the type of bonding that holds the molecules, atoms, or ions together in each solid. Depending on the type of bonding, solids may be described as ionic, molecular, metallic, or covalent network solids. The properties to be studied are relative melting point, solubility in aqueous solution, and electrical conductivity.

OBJECTIVES

• **Observe** the physical properties of different solids.

• **Relate** knowledge of these properties to the type of bonding in each solid.

• **Identify** the type of bonding in an unknown solid.

MATERIALS

- beakers, 50 mL (6)
- Bunsen burner
- copper wire
- deionized water
- evaporating dishes or crucibles (6)
- graduated cylinder, 10 mL
- aluminum shot
- LED conductivity tester
- silicon dioxide (sand)
- sodium chloride (NaCl)

- spatula
- sucrose
- test tubes, small, with solid rubber stoppers (6)
- test-tube rack
- tongs
- unknown substance
- wire gauze, support stand, iron ring, and clay triangle

Always wear safety goggles and a lab apron to protect your eyes and clothing. If you get a chemical in your eyes, immediately flush the chemical out at the eyewash station while calling to your teacher. Know the locations of the emergency lab shower and the eyewash station and the procedures for using them.

Do not touch any chemicals. If you get a chemical on your skin or clothing, wash the chemical off at the sink while calling to your teacher. Make sure you carefully read the labels and follow the precautions on all containers of chemicals that you use. If there are no precautions stated on the label, ask your teacher what precautions you should follow. Do not taste any chemicals or items used in the laboratory. Never return leftovers to their original container; take only small amounts to avoid wasting supplies.

When using a Bunsen burner, confine long hair and loose clothing. If your clothing catches on fire, WALK to the emergency lab shower and use it to put out the fire. When heating a substance in a test tube, the mouth of the test tube should point away from where you and others are standing. Watch the test tube at all times to prevent the contents from boiling over.

Never put broken glass in a regular waste container. Broken glass should be disposed of separately according to your teacher's instructions.

PREPARATION

Use the **Data Table** to record the results of melting, water solubility, solid conductivity, aqueous solution conductivity, and type of bonding of each substance tested.

PROCEDURE

1. Place 1 g samples of each substance into separate evaporating dishes.

2. Touch the electrodes of the conductivity tester to each solid. After each test, rinse with deionized water and carefully dry the electrodes. Note which substances conducted electricity.

3. Place one evaporating dish on a triangle, and heat with a Bunsen burner. As soon as a solid melts, remove the flame.

4. Repeat this procedure for every substance. Do not heat any substance for more than 5 min. There may be some substances that will not melt.

5. Note which substances melted and how long the substances took to melt.

6. Place five test tubes in the test-tube rack. Place 0.5 g of each solid into its own individual test tube. Add 5 mL of deionized water to each test tube. Stopper and shake each test tube in an attempt to dissolve the solid.

7. Note which substances dissolved in the water.

8. Place the solutions or mixtures into separate 50 mL beakers, and immerse the electrodes of the conductivity tester. Rinse the electrodes with the solvent (deionized water) before and after each test. Note which substances conduct electricity.

DISPOSAL

9. Dispose of solids and solutions in containers designated by your teacher.

10. Clean all equipment and return it to its proper place.

11. Wash your hands thoroughly after cleaning up your area and equipment.

Name _____ Class _____ Date _____

Data Table						
	NaCl	**Sucrose**	**Copper wire**	**Iron filings**	**SiO$_2$**	**CaCl$_2$**
Melting	just melts	melts easily	does not melt	does not melt	does not melt	does not melt or just melts
Water solubility	soluble	soluble	insoluble	insoluble	insoluble	soluble
Solid conductivity	no	no	yes	yes	no	no
Aqueous solution (or mixture) conductivity	yes (strong)	no (or weak)	no	no	no	yes
Type of bonding	ionic	molecular	metallic	metallic	covalent network	ionic

Analysis

1. Analyzing Methods Why did you rinse the electrodes before each conductivity test?

The electrodes need to be washed with deionized water before each conductiv-

ity test to ensure that none of the previous material is clinging to the elec-

trodes and thereby giving a false conductivity reading.

_____

_____

_____

2. Analyzing Methods Why did you use deionized water in making the solutions?

Deionized water is used to make the solutions instead of tap water because

tap water contains ions, which could give a false positive conductivity test

result.

_____

_____

_____

Types of Bonding in Solids *continued*

3. Organizing Data List the results that each type of bonding should show.

Ionic bonding shows high melting, solubility in water, no conductivity as a

solid, but conductivity as an aqueous solution. Molecular bonding shows low

melting, solubility in water, no conductivity as a solid or in aqueous solution.

Metallic bonding shows high melting point, no solubility in water, conductiv-

ity as a solid, but no conductivity in aqueous solution. Covalent network

bonding shows high melting point, no solubility in water, no conductivity as

solid or in aqueous solution.

Conclusions

1. Inferring Conclusions What type of bonding describes each substance?
Explain your reasoning.

Sodium chloride is ionic because it has a high melting point, and conducts

electricity when dissolved in water but not in the solid phase. This

conductivity in solution is caused by ionic solids dissolving and forming ions

in solution, which conduct the electricity. Sucrose is molecular because it

has a low melting point and does not conduct electricity in solution or in

the solid state. The low melting point is due to the presence of weak

intermolecular forces that are easily broken when going from solid to liquid.

The lack of conductivity is due to the lack of ions in the solution. Copper

and iron are metallic because they have very high melting points and conduct

electricity in the solid state. This conductivity is due to the easily mobile

valence electrons in the solid. SiO_2 is a covalent network because it has

a very high melting point and does not conduct electricity. Its lack of

conductivity is due to the strength of the covalent bonds (and lack of ions).

It has a very high melting point because the solid contains only covalent

bonds and has no weak intermolecular forces.

Types of Bonding in Solids *continued*

2. Inferring Conclusions Comparing the properties of your unknown solid with the properties of the known solids, determine the type of bonding present in your unknown solid.

Students should identify the unknown as ionic and similar in character to

NaCl.

_____

EXTENSIONS

1. Evaluating Methods Is it possible, for a specific type of bonding, for these properties to vary from what was observed in this experiment? If so, give an example of such a variance.

Discrepancies may arise because of differences in the molecular type of

bonding. If the molecule is nonpolar, then it will not be soluble in aqueous

solution. If the molecule is acidic or basic (forms ions in solution), then it

will show some electrical conductivity in solution. However, this conductivity

will not be as great as in an ionic solution.

_____

_____

2. Applying Conclusions Think about diamond. What would you predict to be the results of this experiment performed on diamond, and what would you predict the bond type to be?

Diamond would not melt, would not be soluble in aqueous solution, and

would not conduct electricity as a solid or mixture. Hence, diamond must

have covalent-network bonding.

_____

_____

Name _____ Class _____ Date _____

Determining the Empirical Formula of Magnesium Oxide

This gravimetric analysis involves the combustion of magnesium metal in air to synthesize magnesium oxide. The mass of the product is greater than the mass of magnesium used because oxygen reacts with the magnesium metal. As in all gravimetric analyses, success depends on attaining a product yield near 100%. Therefore, the product will be heated and cooled and have its mass measured until two of these mass measurements are within 0.02% of one another. When the masses of the reactant and product have been carefully measured, the amount of oxygen used in the reaction can be calculated. The ratio of oxygen to magnesium can then be established, and the empirical formula of magnesium oxide can be determined.

OBJECTIVES

• **Measure** the mass of magnesium oxide.

• **Perform** a synthesis reaction by using gravimetric techniques.

• **Determine** the empirical formula of magnesium oxide.

• **Calculate** the class average and standard deviation for moles of oxygen used.

MATERIALS

• 15 cm magnesium ribbon, 2

• 25 mL beaker

• Bunsen burner assembly

• clay triangle

• crucible and lid, metal or ceramic

• crucible tongs

• distilled water

• eyedropper or micropipet

• ring stand

Always wear safety goggles and a lab apron to protect your eyes and clothing. If you get a chemical in your eyes, immediately flush the chemical out at the eyewash station while calling to your teacher. Know the locations of the emergency lab shower and the eyewash station and the procedures for using them.

When using a Bunsen burner, confine long hair and loose clothing. If your clothing catches on fire, WALK to the emergency lab shower and use it to put out the fire. When heating a substance in a test tube, the mouth of the test tube should point away from where you and others are standing. Watch the test tube at all times to prevent the contents from boiling over.

 Never put broken glass in a regular waste container. Broken glass should be disposed of separately according to your teacher's instructions.

PREPARATION

Use the **Data Table** provided to record your data.

Name _____ Class _____ Date _____

Determining the Empirical Formula of Magnesium Oxide *continued*

PROCEDURE

1. Construct a setup for heating a crucible as demonstrated in the Pre-Laboratory Procedure "Gravimetric Analysis."

2. Heat the crucible and lid for 5 min to burn off any impurities.

3. Cool the crucible and lid to room temperature. Measure their combined mass, and record the measurement on line 3 of the **Data Table.**

   NOTE: Handle the crucible and lid with crucible tongs at all times during this experiment. Such handling prevents burns and the transfer of dirt and oil from your hands to the crucible and lid.

4. Polish a 15 cm strip of magnesium with steel wool. The magnesium should be shiny. Cut the strip into small pieces to make the reaction proceed faster, and place the pieces in the crucible.

5. Cover the crucible with the lid, and measure the mass of the crucible, lid, and metal. Record the measurement on line 1 of the **Data Table.**

6. Use tongs to replace the crucible on the clay triangle. Heat the covered crucible gently. Lift the lid occasionally to allow air in.

   CAUTION: Do not look directly at the burning magnesium metal. The brightness of the light can blind you.

7. When the magnesium appears to be fully reacted, partially remove the crucible lid and continue heating for 1 min.

8. Remove the burner from under the crucible. After the crucible has cooled, use an eyedropper to carefully add a few drops of water to decompose any nitrides that may have formed.

   CAUTION: Use care when adding water. Using too much water can cause the crucible to crack.

9. Cover the crucible completely. Replace the burner under the crucible, and continue heating for about 30 to 60 s.

10. Turn off the burner. Cool the crucible, lid, and contents to room temperature. Measure the mass of the crucible, lid, and product. Record the measurement in the margin of the **Data Table.**

11. Replace the crucible, lid, and contents on the clay triangle, and reheat for another 2 min. Cool to room temperature, and remeasure the mass of the crucible, lid, and contents. Compare this mass measurement with the measurement obtained in step **10.** If the new mass is ±0.02% of the mass in step **10,** record the new mass on line 2 of your data table and go on to step **12.** If not, your reaction is still incomplete, and you should repeat step **11.**

12. Clean the crucible, and repeat steps **2–11** with a second strip of magnesium ribbon. Record your measurements under Trial 2 in the **Data Table.**

DISPOSAL

13. Put the solid magnesium oxide in the designated waste container. Return any unused magnesium ribbon to your teacher. Clean your equipment and lab station. Thoroughly wash your hands after completing the lab session and cleanup.

Data Table	Trial 1	Trial 2
1. Mass of crucible, lid, and metal (g)	6.218	
2. Mass of crucible, lid, and product (g)	6.336	
3. Mass of crucible and lid (g)	6.000	

Analysis

1. Applying Ideas Calculate the mass of the magnesium metal and the mass of the product.

magnesium = 0.218 g; product = 0.336 g

2. Evaluating Data Determine the mass of the oxygen consumed.

oxygen = 0.118 g

3. Applying Ideas Calculate the number of moles of magnesium and the number of moles of oxygen in the product.

moles of oxygen = 0.00738; moles of magnesium = 0.00897

Determining the Empirical Formula of Magnesium Oxide *continued*

Conclusions

1. **Inferring Relationships** Determine the empirical formula for magnesium oxide, Mg_xO_y.

   The mole ratio of Mg to O is 0.00897:0.00738, or 1:0.82. The empirical

   formula is probably MgO.

   _____

   _____

Quick Lab

Balancing Equations Using Models

How can molecular models and formula-unit ionic models be used to balance chemical equations and classify chemical reactions?

MATERIALS

- large and small gumdrops in at least four different colors
- toothpicks

Always wear safety goggles and a lab apron to protect your eyes and clothing. If you get a chemical in your eyes, immediately flush the chemical out at the eyewash station while calling to your teacher. Know the locations of the emergency lab shower and the eyewash station and the procedures for using them.

PROCEDURE

Examine the partial equations in Groups A–E. Using different-colored gumdrops to represent atoms of different elements, make models of the reactions by connecting the appropriate "atoms" with toothpicks. Use your models to (1) balance equations (a) and (b) in each group, (2) determine the products for reaction (c) in each group, and (3) complete and balance each equation (c). Finally, (4) classify each group of reactions by type.

Group A

a. $H_2 + Cl_2 \longrightarrow HCl$

$H_2 + Cl_2 \rightarrow 2HCl$

b. $Mg + O_2 \longrightarrow MgO$

$2Mg + O_2 \rightarrow 2MgO$

c. $BaO + H_2O \longrightarrow$ _____ $Ba(OH)_2$

$BaO + H_2O \rightarrow Ba(OH)_2$

reaction type: _____ synthesis _____

Group B

a. $H_2CO_3 \longrightarrow CO_2 + H_2O$

$H_2CO_3 \rightarrow CO_2 + H_2O$

b. $KClO_3 \longrightarrow KCl + O_2$

$2KClO_3 \rightarrow 2KCl + 3O_2$

c. $H_2O \xrightarrow{electricity}$ _____ $H_2 + O_2$ _____

$2H_2O \rightarrow 2H_2 + O_2$

reaction type: _____ decomposition _____

Balancing Equations Using Models *continued*

Group C

a. $Ca + H_2O \longrightarrow Ca(OH)_2 + H_2$

$\underline{Ca + 2H_2O \rightarrow Ca(OH)_2 + H_2}$

b. $KI + Br_2 \longrightarrow KBr + I_2$

$\underline{2KI + Br_2 \rightarrow 2KBr + I_2}$

c. $Zn + HCl \longrightarrow$ $\underline{\quad H_2 + ZnCl_2 \quad}$

$\underline{Zn + 2HCl \rightarrow ZnCl_2 + H_2}$

reaction type: $\underline{\quad \text{single displacement} \quad}$

Group D

a. $AgNO_3 + NaCl \longrightarrow AgCl + NaNO_3$

$\underline{AgNO_3 + NaCl \rightarrow AgCl + NaNO_3}$

b. $FeS + HCl \longrightarrow FeCl_2 + H_2S$

$\underline{FeS + 2HCl \rightarrow FeCl_2 + H_2S}$

c. $H_2SO_4 + KOH \longrightarrow$ $\underline{\quad K_2SO_4 + H_2O \quad}$

$\underline{H_2SO_4 + 2KOH \rightarrow K_2SO_4 + 2H_2O}$

reaction type: $\underline{\quad \text{double displacement} \quad}$

Group E

a. $CH_4 + O_2 \longrightarrow CO_2 + H_2O$

$\underline{CH_4 + 2O_2 \rightarrow CO_2 + 2H_2O}$

b. $CO + O_2 \longrightarrow CO_2$

$\underline{2CO_2 + O_2 \rightarrow 2CO_2}$

c. $C_3H_8 + O_2 \longrightarrow$ $\underline{\quad CO_2 + H_2O \quad}$

$\underline{C_3H_8 + 5O_2 \rightarrow 3CO_2 + 4H_2O}$

reaction type: $\underline{\quad \text{combustion} \quad}$

Inquiry

Blueprint Paper

Blueprint paper is prepared by coating paper with a solution of two soluble iron(III) salts—potassium hexacyanoferrate(III), commonly called potassium ferricyanide, and iron(III) ammonium citrate. These two salts do not react with each other in the dark. However, when exposed to UV light, the iron(III) ammonium citrate is converted to an iron(II) salt. Potassium hexacyanoferrate(III), $K_3Fe(CN)_6$, reacts with iron(II) ion, Fe^{2+}, to produce an insoluble blue compound, $KFeFe(CN)_6 \cdot H_2O$. In this compound, iron appears to exist in both the +2 and +3 oxidation states.

A blueprint is made by using black ink to make a sketch on a piece of tracing paper or clear, colorless plastic. This sketch is placed on top of a piece of blueprint paper and exposed to ultraviolet light. Wherever the light strikes the paper, the paper turns blue. The paper is then washed to remove the soluble unexposed chemical and is allowed to dry. The result is a blueprint—a blue sheet of paper with white lines.

OBJECTIVE

• **Prepare** blueprint paper and create a blueprint.

MATERIALS

• 10% iron(III) ammonium citrate solution

• 10% potassium hexacyanoferrate(III) solution

• 25 mL graduated cylinders, 2

• corrugated cardboard, 20 cm × 30 cm, 2 pieces

• glass stirring rod

• Petri dish

• thumbtacks, 4

• tongs

• white paper, 8 cm × 15 cm, 1 piece

Always wear safety goggles and a lab apron to protect your eyes and clothing. If you get a chemical in your eyes, immediately flush the chemical out at the eyewash station while calling to your teacher. Know the locations of the emergency lab shower and the eyewash station and the procedures for using them.

Do not touch any chemicals. If you get a chemical on your skin or clothing, wash the chemical off at the sink while calling to your teacher. Make sure you carefully read the labels and follow the precautions on all containers of chemicals that you use. If there are no precautions stated on the label, ask your teacher what precautions you should follow. Do not taste any chemicals or items used in the laboratory. Never return leftovers to their original container; take only small amounts to avoid wasting supplies.

Blueprint Paper *continued*

PROCEDURE

1. Pour 15 mL of a 10% solution of potassium hexacyanoferrate(III) solution into a Petri dish. With most of the classroom lights off or dimmed, add 15 mL of 10% iron(III) ammonium citrate solution. Stir the mixture.

2. Write your name on an 8 cm × 15 cm piece of white paper. Carefully coat one side of the piece of paper by using tongs to drag it over the top of the solution in the Petri dish.

3. With the coated side up, tack your wet paper to a piece of corrugated cardboard, and cover the paper with another piece of cardboard. **Wash your hands before proceeding to step 4.**

4. Take your paper and cardboard assembly outside into the direct sunlight. Remove the top piece of cardboard so that the paper is exposed. Quickly place an object such as a fern, a leaf, or a key on the paper. If it is windy, you may need to put small weights, such as coins, on the object to keep it in place.

5. After about 20 min, remove the object and again cover the paper with the cardboard. Return to the lab, remove the tacks, and thoroughly rinse the blueprint paper under cold running water. Allow the paper to dry. In your notebook, record the amount of time that the paper was exposed to sunlight.

DISPOSAL

6. Clean all equipment and your lab station. Return equipment to its proper place. Dispose of chemicals and solutions in the containers designated by your teacher. Do not pour any chemicals down the drain or in the trash unless your teacher directs you to do so. Wash your hands thoroughly before you leave the lab and after all work is finished.

Analysis

1. Relating Ideas Why is the iron(III) ammonium citrate solution stored in a brown bottle?

Iron(III) ammonium citrate solution is sensitive to light. The brown bottle

filters out the ultraviolet light.

_____

_____

2. Organizing Ideas When iron(III) ammonium citrate is exposed to light, the oxidation state of the iron changes. What is the new oxidation state of the iron?

The new oxidation state is +2.

_____

Blueprint Paper *continued*

3. Analyzing Methods What substances were washed away when you rinsed the blueprint in water after it had been exposed to sunlight? (Hint: Compare the solubilities of the two ammonium salts that you used to coat the paper and of the blue product that formed.)

The two compounds that were mixed and spread on the paper— iron(III)

ammonium citrate and potassium hexacyanoferrate(III)— were washed

away. They are soluble and did not react because the iron(III) ammonium

citrate was not activated by UV radiation.

Conclusions

1. Applying Ideas Insufficient washing of the exposed blueprints results in a slow deterioration of images. Suggest a reason for this deterioration.

If some unreacted chemicals are left on the blueprint, exposure to light over

a period of time can slowly cause a reaction that will adversely affect the

original image.

2. Relating Ideas Photographic paper can be safely exposed to red light in a darkroom. Do you think the same would be true of blueprint paper? Explain your answer.

The wavelength of red light is longer than the wavelength of ultraviolet

light. The less energetic red light does not activate the light-sensitive

chemicals on photographic paper. The same is true for blueprint paper.

Blueprint Paper *continued*

EXTENSIONS

1. **Applying Ideas** How could you use this blueprint paper to test the effectiveness of a brand of sunscreen lotion?

   Students' answers will vary. One possible procedure is to spread a thin layer

   of sunscreen lotion on half of a piece of transparent acetate and place it

   over the blueprint paper before exposing it to sunlight. After developing the

   blueprint, compare the two areas.

   _____

2. **Designing Experiments** Can you think of ways to improve this procedure? If
   so, ask your teacher to approve your plan, and create a new blueprint.
   Evaluate both the efficiency of the procedure and the quality of blueprint.

   Students' answers will vary. Be sure that student procedures are safe.

   _____

   _____

   _____

   _____

   _____

   _____

   _____

DATASHEET FOR IN-TEXT LAB

Limiting Reactants in a Recipe

MATERIALS

- 1/2 cup sugar
- 1/2 cup brown sugar
- 1 1/3 stick margarine (at room temperature)
- 1 egg
- 1/2 tsp. salt
- 1 tsp. vanilla
- 1/2 tsp. baking soda

- 1 1/2 cup flour
- 1 1/3 cup chocolate chips
- mixing bowl
- mixing spoon
- measuring spoons and cups
- cookie sheet
- oven preheated to 350°F

PROCEDURE

1. In the mixing bowl, combine the sugars and margarine together until smooth. (An electric mixer will make this process go much faster.)
2. Add the egg, salt, and vanilla. Mix well.
3. Stir in the baking soda, flour, and chocolate chips. Chill the dough for an hour in the refrigerator for best results.
4. Divide the dough into 24 small balls about 3 cm in diameter. Place the balls on an ungreased cookie sheet.
5. Bake at 350°F for about 10 minutes, or until the cookies are light brown.
   Yield: 24 cookies

DISCUSSION

1. Suppose you are given the following amounts of ingredients:

   1 dozen eggs

   24 tsp. of vanilla

   1 lb. (82 tsp.) of salt

   1 lb. (84 tsp.) of baking soda

   3 cups of chocolate chips

   5 lb. (11 cups) of sugar

   2 lb. (4 cups) of brown sugar

   1 lb. (4 sticks) of margarine

 Teacher's Guide: Stoichiometry

a. For each ingredient, calculate how many cookies could be prepared if all of that ingredient were consumed. (For example, the recipe shows that using 1 egg—with the right amounts of the other ingredients—yields 24 cookies. How many cookies can you make if the recipe is increased proportionately for 12 eggs?)

1 dozen eggs: 288 cookies
_____

24 tsp. of vanilla: 576 cookies
_____

82 tsp. of salt: 3936 cookies
_____

84 tsp. of baking soda: 4032 cookies
_____

3 cups of chocolate chips: 54 cookies
_____

11 cups of sugar: 528 cookies
_____

4 cups of brown sugar: 192 cookies
_____

4 sticks of margarine: 72 cookies
_____

b. To determine the limiting reactant for the new ingredients list, identify which ingredient will result in the fewest number of cookies.

chocolate chips
_____

_____

c. What is the maximum number of cookies that can be produced from the new amounts of ingredients?

54 cookies
_____

_____

Name _____ Class _____ Date _____

Stoichiometry and Gravimetric Analysis

This gravimetric analysis involves a double-displacement reaction between strontium chloride, $SrCl_2$, and sodium carbonate, Na_2CO_3. This type of reaction can be used to determine the amount of a carbonate compound in a solution. For accurate results, essentially all of the reactant of unknown amount must be converted into product. If the mass of the product is carefully measured, you can use stoichiometric calculations to determine how much of the reactant of unknown amount was involved in the reaction.

OBJECTIVES

- **Observe** the double-displacement reaction between solutions of strontium chloride and sodium carbonate.

- **Demonstrate** proficiency with gravimetric methods.

- **Measure** the mass of the precipitate that forms.

- **Relate** the mass of the precipitate that forms to the mass of the reactants before the reaction.

- **Calculate** the mass of sodium carbonate in a solution of unknown concentration.

MATERIALS

- 15 mL Na_2CO_3 solution of unknown concentration
- 50 mL 0.30 M $SrCl_2$ solution
- 50 mL graduated cylinder
- 250 mL beakers, 2
- balance
- beaker tongs
- distilled water
- drying oven

- filter paper
- glass funnel or Büchner funnel with related equipment
- glass stirring rod
- paper towels
- ring and ring stand
- spatula
- water bottle

Always wear safety goggles and a lab apron to protect your eyes and clothing. If you get a chemical in your eyes, immediately flush the chemical out at the eyewash station while calling to your teacher. Know the locations of the emergency lab shower and the eyewash station and the procedures for using them.

Do not touch any chemicals. If you get a chemical on your skin or clothing, wash the chemical off at the sink while calling to your teacher. Make sure you carefully read the labels and follow the precautions on all containers of chemicals that you use. If there are no precautions stated on the label, ask your teacher what precautions you should follow. Do not taste any chemicals or items used in the

| Stoichiometry and Gravimetric Analysis *continued*

laboratory. Never return leftovers to their original container; take only small amounts to avoid wasting supplies.

 Never put broken glass in a regular waste container. Broken glass should be disposed of separately according to your teacher's instructions.

PREPARATION

1. Use the data table provided to record your data.

2. Clean all of the necessary lab equipment with soap and water, and rinse with distilled water.

3. Measure the mass of a piece of filter paper to the nearest 0.01 g, and record it in your table.

4. Set up a filtering apparatus. Use the Pre-Laboratory Procedure "Extraction and Filtration."

5. Label a paper towel with your name and the date. Place the towel in a clean, dry 250 mL beaker, and measure and record the mass of the paper towel and beaker to the nearest 0.01 g.

PROCEDURE

1. Measure about 15 mL of the Na_2CO_3 solution into the graduated cylinder. Record this volume to the nearest 0.5 mL. Pour the Na_2CO_3 solution into an empty 250 mL beaker. Carefully wash the graduated cylinder, and rinse it with distilled water.

2. Measure about 25 mL of the 0.30 M $SrCl_2$ solution into the graduated cylinder. Record this volume to the nearest 0.5 mL. Pour the $SrCl_2$ solution into the beaker with the Na_2CO_3 solution. Gently stir with a glass stirring rod.

3. Measure another 10 mL of the $SrCl_2$ solution into the graduated cylinder. Record the volume to the nearest 0.5 mL. Slowly add the solution to the beaker, and stir gently. Repeat this step until no more precipitate forms.

4. Slowly pour the mixture into the funnel. Do not overfill the funnel—some of the precipitate could be lost between the filter paper and the funnel.

5. Rinse the beaker several more times with distilled water. Pour the rinse water into the funnel each time.

6. After all of the solution and rinses have drained through the funnel, use distilled water to slowly rinse the precipitate on the filter paper in the funnel to remove any soluble impurities.

7. Carefully remove the filter paper from the funnel, and place it on the paper towel that you labeled with your name. Unfold the filter paper, and place the paper towel, filter paper, and precipitate in the rinsed beaker. Then, place the beaker in the drying oven. For best results, allow the precipitate to dry overnight.

| Stoichiometry and Gravimetric Analysis *continued*

8. Using beaker tongs, remove your sample from the oven, and let it cool. Record the total mass of the beaker, paper towel, filter paper, and precipitate to the nearest 0.01 g.

DISPOSAL

9. Dispose of the precipitate and the filtrate in designated waste containers. Clean up all equipment after use, and dispose of substances according to your teacher's instructions. Wash your hands thoroughly after all lab work is finished.

Data Table	
Volume of Na_2CO_3 solution added	**15.0 mL**
Volume of $SrCl_2$ solution added	**35.0 mL**
Mass of dry filter paper	**0.30 g**
Mass of beaker with paper towel	**103.20 g**
Mass of beaker with paper towel, filter paper, and precipitate	**104.56 g**

Analysis

1. Organizing Ideas Write a balanced equation for the reaction. What is the precipitate?

$$SrCl_2(aq) + Na_2CO_3(aq) \rightarrow 2NaCl(aq) + SrCO_3(s)$$

The precipitate is strontium carbonate, $SrCO_3$.

2. Applying Ideas Calculate the mass of the dry precipitate. Calculate the number of moles of precipitate produced in the reaction.

mass of dry $SrCO_3$ = 104.56 g − (103.20 + 0.30)g = 1.06 g $SrCO_3$

mol of $SrCO_3$ = 1.06 g $\times \dfrac{1 \text{ mol } SrCO_3}{147.63 \text{ g}}$ = 7.18 $\times$ 10^{-3} mol $SrCO_3$

Stoichiometry and Gravimetric Analysis *continued*

3. Applying Ideas How many moles of Na_2CO_3 were present in the 15 mL sample? How many grams of Na_2CO_3 were present?

moles of Na_2CO_3 in 15 mL sample =

$$7.18 \times 10^{-3} \text{ mol SrCO}_3 \times \frac{1 \text{ mol Na}_2\text{CO}_3}{1 \text{ mol SrCO}_3} = 7.18 \times 10^{-3} \text{ mol Na}_2\text{CO}_3$$

mass of Na_2CO_3 in 15 mL sample =

$$7.18 \times 10^{-3} \text{ mol Na}_2\text{CO}_3 \times \frac{105.99 \text{ g}}{1 \text{ mol Na}_2\text{CO}_3} = 0.761 \text{ g Na}_2\text{CO}_3/15 \text{ mL}$$

Conclusions

1. Applying Conclusions There are 0.30 mol $SrCl_2$ in every liter of solution. Calculate the number of moles of $SrCl_2$ that were added. What is the limiting reactant?

$$35.0 \text{ mL SrCl}_2 \times \frac{1\text{L}}{1000 \text{ mL}} \times \frac{0.30 \text{ mol}}{1\text{L}} = 1.05 \times 10^{-2} \text{ mol SrCl}_2$$

Sodium carbonate is the limiting reactant.

Name _____   Class _____   Date _____

Microscale

"Wet" Dry Ice

The phase diagram for carbon dioxide shows that CO_2 can exist only as a gas at ordinary room temperature and pressure. To observe the transition of solid CO_2 to liquid CO_2, you must increase the pressure until it is at or above the triple point pressure.

OBJECTIVES

- **Interpret** a phase diagram.
- **Observe** the melting of CO_2 while varying pressure.
- **Relate** observations of CO_2 to its phase diagram.

MATERIALS

- 4–5 g CO_2 as dry ice, broken into rice-sized pieces
- forceps
- metric ruler
- plastic pipets, 5 mL, shatterproof
- pliers
- scissors
- transparent plastic cup

Always wear safety goggles and a lab apron to protect your eyes and clothing. If you get a chemical in your eyes, immediately flush the chemical out at the eyewash station while calling to your teacher. Know the locations of the emergency lab shower and the eyewash station and the procedures for using them.

Do not touch any chemicals. If you get a chemical on your skin or clothing, wash the chemical off at the sink while calling to your teacher. Make sure you carefully read the labels and follow the precautions on all containers of chemicals that you use. If there are no precautions stated on the label, ask your teacher what precautions you should follow. Do not taste any chemicals or items used in the laboratory. Never return leftovers to their original container; take only small amounts to avoid wasting supplies.

PROCEDURE

1. Use forceps to place 2–3 very small pieces of dry ice on the table, and observe them until they have completely sublimed. Record your observations in the space provided. **Caution:** Dry ice will freeze skin very quickly. Do not attempt to pick up the dry ice with your fingers.

2. Fill a plastic cup with tap water to a depth of 4–5 cm.

3. Cut the tapered end (tip) off the graduated pipet.

4. Use forceps to carefully slide 8–10 pieces of dry ice down the stem and into the bulb of the pipet.

5. Use a pair of pliers to clamp the opening of the pipet stem securely shut so that no gas can escape. Use the pliers to hold the tube and to lower the pipet into the cup just until the bulb is submerged. From the side of the cup, observe the behavior of the dry ice. Record your observations in the space provided.

6. As soon as the dry ice has begun to melt, quickly loosen the pliers while still holding the bulb in the water. Observe the CO_2 and record your observations in the space provided.

7. Tighten the pliers again, and record your observations.

8. Repeat Procedure steps 6 and 7 as many times as possible.

DISPOSAL

9. Clean all apparatus and your lab station. Return equipment to its proper place. Dispose of chemicals and solutions in the containers designated by your teacher. Do not pour any chemicals down the drain or place them in the trash unless your teacher directs you to do so. Wash your hands thoroughly before you leave the lab and after all work is finished.

OBSERVATIONS

Procedure 1 Student answers will vary, but should include a white "mist" coming from the dry ice, the "mist" hoves on the table top, and the pieces eventually vanish.

_____

Procedure 5 Student answers will vary, but the dry ice should show signs of melting.

_____

Procedure 6 and 7 Student answers will vary, but the melted dry ice should solidify.

_____

_____

Name _____ Class _____ Date _____

"Wet" Dry Ice continued

Analysis

1. **Analyzing Results** What differences did you observe between the subliming and the melting of CO_2?

   Students' answers may vary. At first, the dry ice sublimes, but then it should

   gradually begin to melt.

   _____

2. **Analyzing Methods** As you melted the CO_2 sample over and over, why did it eventually disappear? What could you have done to make the sample last longer?

   The sample was used up because every time the pliers were loosened, some

   gaseous CO_2 escaped.

   _____

3. **Analyzing Methods** What purpose(s) do you suppose the water in the cup served?

   Students' answers will vary but may include one or more of the following:

   The water in the cup provided the energy as heat that the dry ice needed to

   absorb in order to sublime and melt. This was important because plastic

   materials like the pipet bulb tend to get brittle when cold. The water also

   absorbed the energy when the liquid CO_2 froze again, so the temperature of

   the system did not fluctuate much. If the experiment had been performed

   outside the water, condensation would have formed on the outside of the

   pipet bulb, obscuring the view. Because the cup is curved, the water slightly

   magnified the view of the dry ice in the pipet bulb. Finally, if the bulb had

   burst because of the pressure of CO_2, the water would have caught the dry

   ice and the pieces of the bulb and would have muffled the sound of the small

   explosion.

Name _____ Class _____ Date _____

"Wet" Dry Ice *continued*

EXTENSIONS

1. **Predicting Outcomes** What would have happened if fewer pieces of dry ice (only 1 or 2) had been placed inside the pipet bulb? If time permits, test your prediction.

   If fewer pieces of dry ice were used, it would take longer to reach a pressure

   high enough for the dry ice to melt, if it melted at all.

   _____

   _____

   _____

2. **Predicting Outcomes** What might have happened if too much dry ice (20 or 30 pieces, for example) had been placed inside the pipet bulb? How quickly would the process have occurred? If time permits, test your prediction.

   If more dry ice were used, the conditions for melting would be achieved very

   quickly, but the rapid increase in the pressure in the bulb might cause the bulb

   to stretch or shatter. As this large amount of dry ice sublimed and melted,

   it could also absorb enough energy as heat from the water to turn the water

   into ice.

   _____

   _____

3. **Predicting Outcomes** What would have happened if the pliers had not been released once the dry ice melted? If time permits, test your prediction.

   As the dry ice sublimes and melts, the pressure will continue to build until

   the bulb stretches out or blows apart, and releases the CO_2 gas.

   _____

   _____

Quick Lab

Diffusion

Do different gases diffuse at different rates?

MATERIALS

- household ammonia
- perfume or cologne
- two 250 mL beakers
- two watch glasses
- 10 mL graduated cylinder
- clock or watch with second hand

Always wear safety goggles and a lab apron to protect your eyes and clothing. If you get a chemical in your eyes, immediately flush the chemical out at the eyewash station while calling to your teacher. Know the locations of the emergency lab shower and the eyewash station and the procedures for using them.

Do not touch any chemicals. If you get a chemical on your skin or clothing, wash the chemical off at the sink while calling to your teacher. Make sure you carefully read the labels and follow the precautions on all containers of chemicals that you use. If there are no precautions stated on the label, ask your teacher what precautions you should follow. Do not taste any chemicals or items used in the laboratory. Never return leftovers to their original container; take only small amounts to avoid wasting supplies.

PROCEDURE

Record all of your results in the **Data Table.**

1. Outdoors or in a room separate from the one in which you will carry out the rest of the investigation, pour approximately 10 mL of the household ammonia into one of the 250 mL beakers, and cover it with a watch glass. Pour roughly the same amount of perfume or cologne into the second beaker. Cover it with a watch glass also.

2. Take the two samples you just prepared into a large, draft-free room. Place the samples about 12 to 15 ft apart and at the same height. Position someone as the observer midway between the two beakers. Remove both watch glass covers at the same time.

3. Note whether the observer smells the ammonia or the perfume first. Record how long this takes. Also, record how long it takes the vapor of the other substance to reach the observer. Air the room after you have finished.

DISPOSAL

4. Check with your teacher for the proper disposal procedures. Always wash your hands thoroughly after cleaning up the lab area and equipment.

Data Table

	Ammonia	Perfume/cologne
Location	Answer may vary	Answer may vary
Distance	Answer may vary	Answer may vary
Time (s)	Answer may vary	Answer may vary

DISCUSSION

1. What do the times that the two vapors took to reach the observer show about the two gases?

Ammonia molecules diffuse more rapidly than perfume-scent molecules do.

_____

2. What factors other than molecular mass (which determines diffusion rate) could affect how quickly the observer smells each vapor?

concentration, vapor pressure, and the sensitivity of the observer to the two

odors

_____

_____

Name _____ Class _____ Date _____

Microscale

Mass and Density of Air at Different Pressures

You have learned that the amount of gas present, the volume of the gas, the temperature of the gas sample, and the gas pressure are related to one another. If the volume and temperature of a gas sample are held constant, the mass of the gas and the pressure that the gas exerts are related in a simple way.

In this investigation, you will use an automobile tire pressure gauge to measure the mass of a bottle and the air that the bottle contains for several air pressures. A tire pressure gauge measures "gauge pressure," meaning the added pressure in the tire in addition to normal atmospheric air pressure. Gauge pressure is often expressed in the units pounds per square inch, gauge (psig) to distinguish them from absolute pressures in pounds per square inch (psi). You will graph the mass of the bottle plus air against the gas pressure and observe what kind of plot results. Extrapolating this plot in the proper way will let you determine both the mass and the volume of the empty bottle. This information will also allow you to calculate the density of air at various pressures.

OBJECTIVES

- **Measure** the pressure exerted by a gas.
- **Measure** the mass of a gas sample at different pressures.
- **Graph** the relationship between the mass and pressure of a gas sample.
- **Calculate** the mass of an evacuated bottle.
- **Calculate** the volume of a bottle.
- **Calculate** the density of air at different pressures.

MATERIALS

- automobile tire valve
- balance, centigram
- barometer
- cloth towel
- plastic soda bottle (2 or 3 L) or other heavy plastic bottle
- tire pressure gauge

Always wear safety goggles and a lab apron to protect your eyes and clothing. If you get a chemical in your eyes, immediately flush the chemical out at the eyewash station while calling to your teacher. Know the locations of the emergency lab shower and the eyewash station and the procedures for using them.

PREPARATION

Use the **Data Table** provided to record your data.

PROCEDURE

Your teacher will provide you with a bottle. This bottle contains air under considerable pressure, so handle it carefully. Do not unscrew the cap of the bottle.

1. Use the tire pressure gauge to measure the gauge pressure of the air in the bottle, as accurately as you can read the gauge. It might be convenient for one student to hold the bottle securely, wrapped in a cloth towel, while another student makes the pressure measurement. Record this pressure in the **Data Table.**

2. Measure the mass of the bottle plus the air it contains, to the nearest 0.01 g. Record this mass in the **Data Table.**

3. With one student holding the wrapped bottle, depress the tire stem valve carefully to allow some air to escape from the bottle until the observed gauge pressure has decreased by 5 to 10 psig. Then, repeat the measurements in steps **1** and **2.**

4. Repeat the steps of releasing some pressure (step **3**) and then measuring gauge pressure (step **1**) and measuring the mass (step **2**) until no more air comes out.

5. Now repeat steps **1** and **2** one last time. The gauge pressure should be zero; if it is not, then you probably have not released enough air, and you should depress the valve for a longer time. You should have at least five measurements of gauge pressure and mass, including this final set.

6. Read the atmospheric pressure in the room from the barometer, and record the reading in the **Data Table.**

DISPOSAL

7. Return all equipment to its proper place. Wash your hands thoroughly before you leave the lab and after all work is finished.

Data Table

Gauge Pressure (psig)	Mass of Bottle + Air (g)	Corrected Gas Pressure (psi)	Mass of Air (g)	Density of Air (g/cm³)
0.0	188.70	14.6	3.7	1.2
5.0	189.97	19.6	5.0	1.6
10.0	191.24	24.6	6.2	2.0
15.0	192.51	29.6	7.5	2.4
20.0	193.78	24.6	8.8	2.8
25.0	195.05	39.6	10.0	3.2
30.0	196.32	44.6	11.3	3.7
35.0	197.59	49.6	12.6	4.1
40.0	198.85	54.6	13.8	4.5

Analysis

1. **Organizing Data** Correct each gauge pressure in your data table to the actual gas pressure in psi, by adding the barometric pressure (in psi) to each measured gauge pressure. Enter these results in the column "Corrected gas pressure."

   The following is a representative correction of gauge pressure to gas pressure illustrated for a barometric pressure of 755 torr or 14.6 psi: corrected pressure (psi) = 30.0 psig + 14.6 psi = 44.6 psi

2. **Analyzing Data** Make a graph of your data. Plot corrected gas pressure on the x-axis and mass of bottle plus air on the *y*-axis. The *x*-axis should run from 10 psi to at least 80 psi. The *y*-axis scale should allow extrapolation to corrected gas pressure of zero.

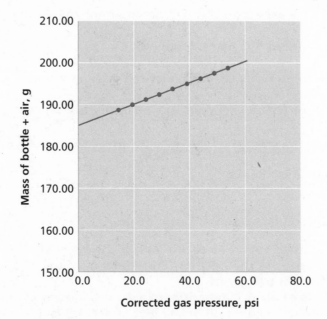

   Above is a representative graph. The sample data plotted here correspond to a nominal three liter bottle with evacuated mass of 185.0 g.

Mass and Density of Air at Different Pressures *continued*

3. Analyzing Data If your graph is a straight line, write an equation for the line in the form $y = mx + b$. If the graph is not a straight line, explain why, and draw the straight line that comes closest to including all of your data points. Give the equation of this line.

The plot should give a straight line. Equations for the line will vary, depending on the volume and mass of the bottle. For the sample data, the equation is

(mass of bottle + air) = 0.254(gas pressure) + 185.0

4. Interpreting Data What is the mass of the empty bottle? (Hint: When no more air escapes from the bottle in steps **4** and **5,** the bottle is not empty; it still contains air at 1 atm.)

Some students may incorrectly assume this to be the measured mass at zero

gauge pressure. At those conditions, however, the bottle still contains air

at one atmosphere pressure. The correct mass for the empty (i.e., evacuated)

bottle can be deduced by one of two methods: (a) the graph can be extra-

polated to 0 corrected gas pressure and the mass of the empty bottle read as

the y-intercept; or (b) the straight-line equation can be evaluated with "gas

pressure" set to 0, corresponding to no air present. For the sample data, the

mass of the bottle is 185.0 g.

5. Analyzing Data For each of your readings, calculate the mass of air in the bottle. Enter these masses in the **Data Table.**

Student answers will vary. For each reading, the mass of air is found by subtracting the mass of the bottle (determined in Analysis Step 6) from the measured mass of bottle + air:

mass of air = (mass of bottle + air) − mass of bottle

For the sample data reading of 196.32 g at 30.0 psig,

Mass of air = 196.32 g − 185.0 g = 11.3 g of air

Mass and Density of Air at Different Pressures *continued*

6. Interpreting Data The density of air at typical laboratory conditions is 1.19 g/L. Find the volume of the bottle.

The value of 1.19 g/L for air density at laboratory conditions is good to within less than 1% at pressures near 1 atm, temperatures of 20–25°C, and over reasonable ranges of relative humidity (20–90%). The volume of the bottle can be found by dividing the determined mass of the air at *atmospheric pressure* (i.e., at gauge pressure of 0) by the density of air. For the sample data (for which mass of air at 0 gauge pressure is 188.70 g − 185.0 g = 3.7 g),

$$\text{volume} = \frac{3.7 \text{ g}}{1.19 \text{ g/L}} = 3.1 \text{ L}$$

7. Interpreting Data Calculate the density of air at each pressure for which you made measurements. Enter these density values in your data table.

For each reading, the density of air is the mass of air divided by the volume of the bottle. For the sample data at gauge pressure 30.0 psig (mass air = 11.3 g),

$$\text{density} = \frac{11.3 \text{ g}}{3.1 \text{ L}} = 3.6 \text{ g/L}$$

Student values should typically range from 1.19 g/L at atmospheric pressure to 5.3 g/L at gauge pressure 50.0 psig.

Conclusions

1. Inferring Relationships Based on your results in this experiment, state the relationship between the mass of a gas sample and the gas pressure. Be sure to include limitations (that is, the quantities that must be kept constant).

At the same volume and temperature, the mass of a gas sample is directly

proportional to the gas pressure.

2. Interpreting Graphics Using your graph from item 2 of Analysis, predict the mass of the bottle plus air at a gauge reading of 60.0 psig. Estimate the mass of the gas in the bottle at that pressure.

Student answers may vary. For the sample data, a gauge pressure of 60.0 psig

gives a corrected pressure of 74.6 psi. Reading the graph for the sample

data gives a value of about 204 g. This would correspond to about

(204 − 185) = 29 grams of gas.

Name _____ Class _____ Date _____

Quick Lab

Observing Solutions, Suspensions, and Colloids

MATERIALS

- balance
- 7 beakers, 400 mL
- clay
- cooking oil
- flashlight
- gelatin, plain
- hot plate (to boil H_2O)
- red food coloring

- sodium borate ($Na_2B_4O_7 \cdot 10H_2O$)
- soluble starch
- stirring rod
- sucrose
- test-tubes, 7
- test-tube rack
- water

  Always wear safety goggles and a lab apron to protect your eyes and clothing. If you get a chemical in your eyes, immediately flush the chemical out at the eyewash station while calling to your teacher. Know the locations of the emergency lab shower and the eyewash station and the procedures for using them.

PROCEDURE

1. Prepare seven mixtures, each containing 250 mL of water and one of the following substances.

   a. 12 g of sucrose

   b. 3 g of soluble starch

   c. 5 g of clay

   d. 2 mL of food coloring

   e. 2 g of sodium borate

   f. 50 mL of cooking oil

   g. 3 g of gelatin

   Making the gelatin mixture: Soften the gelatin in 65 mL of cold water, and then add 185 mL of boiling water.

2. Observe the seven mixtures and their characteristics. Record the appearance of each mixture after stirring.

3. Transfer to individual test tubes 10 mL of each mixture that does not separate after stirring. Shine a flashlight on each mixture in a dark room. Make note of the mixtures in which the path of the light beam is visible.

DISCUSSION

1. Using your observations, classify each mixture as a solution, suspension, or colloid.

sucrose in water: solution; starch in water: colloid; clay in water: suspension;

food coloring in water: solution; sodium borate in water: solution; cooking

oil in water: suspension; gelatin in water: colloid

2. What characteristics did you use to classify each mixture?

If a mixture is cloudy or displays the Tyndall effect, then it is either a colloid

or a suspension. A suspension can be identified by its ability to settle out.

Name _____ Class _____ Date _____

Separation of Pen Inks by Paper Chromatography

Paper Chromatography

Details on this technique can be found in the Pre-Laboratory Procedure "Paper Chromatography" on page 848.

Writing Inks

Most ballpoint pen inks are complex mixtures, containing pigments or dyes that can be separated by paper chromatography.

Black inks can contain three or more colors; the number of colors depends on the manufacturer. Each ink formulation has a characteristic pattern that uniquely identifies it.

In this experiment you will develop radial paper chromatograms for four black ballpoint pen inks, using water as solvent. You will then repeat this process using isopropanol as the solvent. You will then measure the distance traveled by each of the individual ink components and the distance traveled by the solvent front. Finally, you will use these measurements to calculate the R_f factor for each component.

OBJECTIVES

- **Demonstrate** proficiency in qualitatively separating mixtures using paper chromatography.

- **Determine** the R_f factor(s) for each component of each tested ink.

- **Explain** how the inks are separated by paper chromatography.

- **Observe** the separation of a mixture by the method of paper chromatography.

MATERIALS

- 12 cm circular chromatography paper or filter paper, 2

- distilled water

- 1 filter paper wick, 2 cm equilateral triangle

- isopropanol

- numbered pens, each with a different black ink, 4

- pencil

- petri dish with lid

- scissors

Always wear safety goggles and a lab apron to protect your eyes and clothing. If you get a chemical in your eyes, immediately flush the chemical out at the eyewash station while calling to your teacher. Know the locations of the emergency lab shower and the eyewash station and the procedures for using them.

Do not touch any chemicals. If you get a chemical on your skin or clothing, wash the chemical off at the sink while calling to your teacher. Make sure you carefully read the labels and follow the precautions on all containers of chemicals that you use. If there are no precautions stated on the label, ask your teacher what precautions you should follow. Do not taste any chemicals or items used in the laboratory. Never return leftovers to their original container; take only small amounts to avoid wasting supplies.

PREPARATION

1. Determine the formula, structure, polarity, density, and volatility at room temperature for water and isopropanol. The following titles are sources that provide general information on specific elements and compounds: *CRC Handbook of Chemistry and Physics*, *McGraw-Hill Dictionary of Chemical Terms*, and *Merck Index*.

2. Use the data tables provided to record your data.

PROCEDURE

Part A: Prepare a chromatogram using water as the solvent

1. Construct an apparatus for paper chromatography as described in the Pre-Laboratory Procedure on page 848. You will make only four dots. You will use ballpoint pens rather than micropipets to spot your paper.

2. After 15 min or when the water is about 1 cm from the outside edge of the paper, remove the paper from the Petri dish and allow the chromatogram to dry. Record in **Data Table 1** the colors that have separated from each of the four different black inks.

Part B: Prepare a chromatogram using isopropanol as the solvent

3. Repeat Procedure steps **1** to **2**, replacing the water in the Petri dish with isopropanol. Record in **Data Table 2** the colors that have separated from each of the four different black inks.

Part C: Determine R$_f$ values for each component

4. After the chromatogram is dry, use a pencil to mark the point where the solvent front stopped.

5. With a ruler, measure the distance from the initial ink spot to your mark, and record this distance on the appropriate data table.

6. Make a small dot with your pencil in the center of each color band.

7. With a ruler, measure the distance from the initial ink spot to each dot separately, and record each distance on the appropriate data table.

Separation of Pen Inks by Paper Chromatography *continued*

8. Divide each value recorded in Procedure step **7** by the value recorded in Procedure step **5**. The result is the R_f value for that component. Record the R_f values in the appropriate data table. Tape or staple the chromatogram to the appropriate data table.

DISPOSAL

9. The water may be poured down the sink. Chromatograms and other pieces of filter paper may be discarded in the trash. The isopropanol solution should be placed in the waste disposal container designated by your teacher. Clean up your equipment and lab station. Thoroughly wash your hands after completing the lab session and cleanup.

Pen no.	Dot no.	Color 1		Color 2		Color 3		Color 4	
		Distance	R_f value	Distance	R_f value	Distance	R_f value	Distance	R_f value
*	*	*	*	*	*	*	*	*	*

DATA TABLE 1   Chromatogram Formed with Water

Pen no.	Dot no.	Color 1		Color 2		Color 3		Color 4	
		Distance	R_f value	Distance	R_f value	Distance	R_f value	Distance	R_f value
*	*	*	*	*	*	*	*	*	*

DATA TABLE 2   Chromatogram Formed with Isopropanol

* Answers may vary.

Analysis

1. Evaluating Conclusions Is the color in each pen the result of a single dye or multiple dyes? Justify your answer.

The black inks are made with multiple dyes because the inks separate into

more than one color. NOTE: Some inks might not separate when water is the

solvent but should separate when isopropanol is the solvent. Some inks may

separate well with water but may be too soluble in the isopropanol to

separate cleanly.

_____

_____

_____

_____

Separation of Pen Inks by Paper Chromatography *continued*

2. **Relating Ideas** What can be said about the properties of a component ink that has an R_f value of 0.50?

The component is soluble in the solvent but is more attracted to the filter

paper than to the solvent because it moved only half as far.

_____

3. **Analyzing Methods** Suggest a reason for stopping the process when the solvent front is 1 cm from the edge of the filter paper rather than when it is even with the edge of the paper.

Student answers will vary depending on the pens used.

_____

_____

4. **Predicting Outcomes** Predict the results of forgetting to remove the chromatogram from the water in the petri dish until the next day.

If the process continued overnight, the solvent would reach the edge of the

filter paper and begin to evaporate. The slower components would catch up

with the faster ones, and they all would end up near the edge of the filter

paper.

_____

Conclusions

1. **Analyzing Results** Compare the R_f values for the colors from pen number 2 when water was the solvent and the R_f values obtained when isopropanol was the solvent. Explain why they differ.

Student answers will vary but should attribute any differences in the R_f

values to differences in the solvents and their interaction with solutes.

_____

2. **Evaluating Methods** Would you consider isopropanol a better choice for the solvent than water? Why or why not?

Student answers should be based on the quality of the separations. Usually,

isopropanol gives better separation.

_____

Name _____ Class _____ Date _____

Separation of Pen Inks by Paper Chromatography *continued*

3. Analyzing Conclusions Are the properties of the component that traveled the farthest in the water chromatogram likely to be similar to the properties of the component that traveled the farthest in the isopropanol chromatogram? Explain your reasoning.

The component that travels the farthest in the water chromatogram is likely

to be either small and ionic or polar. The component that travels the farthest

in the isopropanol chromatogram is also likely to be small, but probably

nonpolar.

_____

_____

4. Inferring Conclusions What can you conclude about the composition of the inks in ballpoint pens from your chromatogram?

Student answers will vary, but they should indicate that black inks are made

with more than one color of dye and that the number and colors of the dyes

vary. Answers may also include statements about the oily base of the inks if

one or more inks separated in alcohol but not in water or about the glycol

base of the inks if one or more inks separated in water but followed the

solvent front in the isopropanol. Movement through the filter paper depends

on the attraction for both the paper and for the solvent. The more soluble

the component is, the less attracted it is to the paper.

_____

_____

_____

_____

_____

_____

_____

_____

Name _____ Class _____ Date _____

Microscale

Testing Water for Ions

The physical and chemical properties of aqueous solutions are affected by small amounts of dissolved ions. For example, if a water sample has enough Mg^{2+} or Ca^{2+} ions, it does not create lather when soap is added. This is common in places where there are many minerals in the water (hard water). Other ions, such as Pb^{2+} and Co^{2+}, can accumulate in body tissues; therefore, solutions of these ions are poisonous.

Because some sources of water may contain harmful or unwanted substances, it is important to find out what ions are present. In this experiment, you will test various water samples for the presence of four ions: Fe^{3+}, Ca^{2+}, Cl^-, and SO_4^{2-}. Some of the samples may contain these ions in very small concentrations, so make very careful observations.

OBJECTIVES

• **Observe** chemical reactions involving aqueous solutions of ions.

• **Relate** observations of chemical properties to the presence of ions.

• **Infer** whether an ion is present in a water sample.

• **Apply** concepts concerning aqueous solutions of ions.

MATERIALS

• 24-well microplate lid

• fine-tipped dropper bulbs, labeled, with solutions, 10

• overhead projector (optional)

• paper towels

• solution 1: reference (all ions)

• solution 2: distilled water (no ions)

• solution 3: tap water (may have ions)

• solution 4: bottled spring water (may have ions)

• solution 5: local river or lake water (may have ions)

• solution 6: solution X, prepared by your teacher (may have ions)

• solution A: NaSCN solution (test for Fe^{3+})

• solution B: $Na_2C_2O_4$ solution (test for Ca^{2+})

• solution C: $AgNO_3$ solution (test for Cl^-)

• solution D: $Sr(NO_3)_2$ solution (test for SO_4^{2-})

• white paper

◆ ◆ **Always wear safety goggles and a lab apron to protect your eyes and clothing.** If you get a chemical in your eyes, immediately flush the chemical out at the eyewash station while calling to your teacher. Know the locations of the emergency lab shower and the eyewash station and the procedures for using them.

Testing Water for Ions *continued*

Do not touch any chemicals. If you get a chemical on your skin or clothing, wash the chemical off at the sink while calling to your teacher. Make sure you carefully read the labels and follow the precautions on all containers of chemicals that you use. If there are no precautions stated on the label, ask your teacher what precautions you should follow. Do not taste any chemicals or items used in the laboratory. Never return leftovers to their original container; take only small amounts to avoid wasting supplies.

PREPARATION

1. Use the **Data Table** provided to record your observations.

2. Place the 24-well microplate lid in front of you on a white background. Label the columns and rows as instructed by your teacher. The coordinates will designate the individual circles. For example, the circle in the top right corner would be 1-D.

PROCEDURE

1. Obtain labeled dropper bulbs containing the six different solutions from your teacher.

2. Place a drop of the solution from bulb 1 into circles 1-A, 1-B, 1-C, and 1-D (the top row). Solution 1 contains all four of the dissolved ions, so these drops will show what a **positive** test for each ion looks like. **Be careful to keep the solutions in the appropriate circles. Any spills will cause poor results.**

3. Place a drop of the solution from bulb 2 into each of the circles in row 2. This solution is distilled water and should not contain any of the ions. It will show what a **negative** test looks like.

4. Place a drop from bulb 3 into each of the circles in row 3 and a drop from bulb 4 into each of the circles in row 4. Follow the same procedure for bulb 5 (into row 5) and bulb 6 (into row 6). These solutions may or may not contain ions. The materials list gives contents of each bulb.

5. Now that each circle contains a solution to be analyzed, use the solutions in bulbs A–D to test for the presence of the ions. Bulb A contains NaSCN, sodium thiocyanate, which reacts with any Fe^{3+} to form the complex ion $Fe(SCN)^{2+}$, which results in a deep red solution. Bulb B contains $Na_2C_2O_4$, sodium oxalate, which reacts with Ca^{2+} ions. Bulb C contains $AgNO_3$, silver nitrate, which reacts with Cl^- ions. Bulb D contains $Sr(NO_3)_2$, strontium nitrate, which reacts with SO_4^{2-} ions. The contents of bulbs B–D react with the specified ion to yield insoluble precipitates.

6. **Holding the tip of bulb A 1 to 2 cm above the drop of water to be tested,** add one drop of solution A to the drop of reference solution in circle 1-A and one drop to the distilled water in circle 2-A. Circle 1-A should show a positive test, and circle 2-A should show a negative test. In the **Data Table,** record your observations about what the positive and negative tests look like.

7. Use the NaSCN solution in bulb A to test the rest of the water drops in column A to determine whether they contain the Fe^{3+} ion. Record your observations in the **Data Table.** For each of the tests in which the ion was present, specify whether it seemed to be at a high, moderate, or low concentration.

8. Follow the procedure used for bulb A with bulbs B, C, and D to test for the other ions. Record your observations about the test results. Specify whether the solutions contained Ca^{2+}, Cl^-, or SO_4^{2-} and whether the ions seemed to be present at a high, moderate, or low concentration. A black background may be useful for these three tests.

9. If some of the results are difficult to discern, place your microplate on an overhead projector. Examine the drops for signs of cloudiness. Looking at the drops from the side, keep your line of vision 10° to 15° above the plane of the lid. Compare each drop tested with the control drops in row 2. If any sign of cloudiness is detected in a test sample, it is due to the Tyndall effect and is a positive test result. Record your results.

DISPOSAL

10. Clean all equipment and your lab station. Return equipment to its proper place. Dispose of chemicals and solutions in the containers designated by your teacher. Do not pour any chemicals down the drain or in the trash unless your teacher directs you to do so. Wash your hands thoroughly before you leave the lab and after all work is finished.

DATA TABLE				
Test for:	Fe^{3+}	Ca^{2+}	Cl^-	SO_4^{2-}
Reacting with:	SCN^-	$C_2O_4^{2-}$	Ag^+	Sr^{2+}
Reference (all four ions)	red	white	white	white
Distilled H_2O (control—no ions)	colorless	colorless	colorless	colorless
Tap water	*	*	*	*
Bottled spring water	*	*	*	*
River or lake water	*	*	*	*
Solution X	**	**	**	**

* **Observations will vary depending on the source.**

** **Observations will vary depending on the ions selected by teacher.**

Testing Water for Ions *continued*

Analysis

1. **Organizing Ideas** Describe what each positive test looked like. Write the balanced chemical equations and net ionic equations for each of the positive tests.

Test A: red color

$Fe^{3+}(aq) + NaSCN(aq) \rightarrow Na^+(aq) + Fe(SCN)^{2+}(aq)$

$Fe^{3+}(aq) + SCN^-(aq) \rightarrow Fe(SCN)^{2+}(aq)$

Test B: precipitate

$Ca^{2+}(aq) + Na_2C_2O_4(aq) \rightarrow 2Na^+(aq) + CaC_2O_4(s)$

$Ca^{2+}(aq) + C_2O_4^{2-}(aq) \rightarrow CaC_2O_4(s)$

Test C: precipitate

$Cl^-(aq) + AgNO_3(aq) \rightarrow NO_3^-(aq) + AgCl(s)$

$Cl^-(aq) + Ag^+(aq) \rightarrow AgCl(s)$

Test D: precipitate

$SO_4^{2-}(aq) + Sr(NO_3)_2(aq) \rightarrow 2NO_3^-(aq) + SrSO_4(s)$

$SO_4^{2-}(aq) + Sr^{2+}(aq) \rightarrow SrSO_4(s)$

Conclusions

1. **Organizing Conclusions** List the solutions that you tested and the ions that you found in each solution. Include notes on whether the concentration of each ion was high, moderate, or low based on your observations.

Student answers will vary depending on the source of the water samples.

All students should indicate that the reference solution had all of the ions

present and the distilled water had none of the ions present.

Testing Water for Ions *continued*

2. **Predicting Outcomes** Using your test results, predict which water sample would be the "hardest." Explain your reasoning.

Students' answers will vary, depending on the sources of the water samples.

The solution that showed the strongest test for calcium ions is likely to be

"harder" than the other samples because calcium ions are one cause of

hardness in water. Some students may suggest that unless the solutions are

tested for the presence of magnesium ions, you cannot be certain which is

the hardest.

_____

_____

Quick Lab

Household Acids and Bases

Which of the household substances are acids, and which are bases?

MATERIALS

- dishwashing liquid, dishwasher detergent, laundry detergent, laundry stain remover, fabric softener, and bleach
- mayonnaise, baking powder, baking soda, white vinegar, cider vinegar, lemon juice, soft drinks, mineral water, and milk

- fresh red cabbage
- hot plate
- beaker, 500 mL or larger
- beakers, 50 mL
- spatula
- tap water
- tongs

Always wear safety goggles and a lab apron to protect your eyes and clothing. If you get a chemical in your eyes, immediately flush the chemical out at the eyewash station while calling to your teacher. Know the locations of the emergency lab shower and the eyewash station and the procedures for using them.

Do not touch any chemicals. If you get a chemical on your skin or clothing, wash the chemical off at the sink while calling to your teacher. Make sure you carefully read the labels and follow the precautions on all containers of chemicals that you use. If there are no precautions stated on the label, ask your teacher what precautions you should follow. Do not taste any chemicals or items used in the laboratory. Never return leftovers to their original container; take only small amounts to avoid wasting supplies.

Do not heat glassware that is broken, chipped, or cracked. Use tongs or a hot mitt to handle heated glassware and other equipment because hot glassware does not always look hot.

PROCEDURE

Record all your results in the data table provided.

1. To make an acid-base indicator, extract juice from red cabbage. First, cut up some red cabbage and place it in a large beaker. Add enough water so that the beaker is half full. Then, bring the mixture to a boil. Let it cool, and then pour off and save the cabbage juice. This solution is an acid-base indicator.

2. Assemble foods, beverages, and cleaning products to be tested.

3. If the substance being tested is a liquid, pour about 5 mL into a small beaker. If it is a solid, place a small amount into a beaker, and moisten it with about 5 mL of water.

4. Add a drop or two of the red cabbage juice to the solution being tested, and note the color. The solution will turn red if it is acidic and green if it is basic.

DISPOSAL

5. Place all solids in the trash. Pour all liquids down the drain.

Data Table			
Substance	**Color with cabbage juice**	**Substance**	**Color with cabbage juice**
dishwashing liquid	**blue-green to greenish-yellow**	mayonnaise	**red**
dishwashing detergent	**blue-green to greenish-yellow**	baking powder	**blue-violet**
laundry detergent	**blue-green to greenish-yellow**	baking soda	**blue**
laundry stain remover	**blue-green to greenish-yellow**	white vinegar	**red**
fabric softener	**violet-blue**	cider vinegar	**red**
bleach	**greenish-yellow**	lemon juice	**red**
		soft drink	**red**
		mineral water	**purple to violet**
		milk	**purple**

DISCUSSION

1. Are the cleaning products acids, bases, or neither?

Most cleaning products tend to be basic or neutral.

2. What are acid/base characteristics of foods and beverages?

Acidic food products taste tart or sour. Basic food products taste bitter.

Of the food products listed, only baking soda is basic.

3. Did you find consumer warning labels on basic or acidic products?

Consumer warning labels about the acidity or basicity of food products will

probably not be found. Most cleaning products have warning labels; a caustic

warning indicates a strongly basic substance that can cause burns.

Inquiry

Is It an Acid or a Base?

When scientists uncover a problem they need to solve, they think carefully about the problem and then use their knowledge and experience to develop a plan for solving it. In this experiment, you will be given a set of eight colorless solutions. Four of them are acidic solutions (dilute hydrochloric acid) and four are basic solutions (dilute sodium hydroxide). The concentrations of both the acidic and the basic solutions are 0.1 M, 0.2 M, 0.4 M, and 0.8 M. Phenolphthalein has been added to the acidic solutions.

You will first write a procedure to determine which solutions are acidic and which are basic and then carry out your procedure. You will then develop and carry out another procedure that will allow you to order the acidic and basic solutions from lowest to highest concentration. As you plan your procedures, consider the properties of acids and bases that are discussed in Chapter 14. Predict what will happen to a solution of each type and concentration when you do each test. Then compare your predictions with what actually happens. You will have limited amounts of the unknown solutions to work with, so use them carefully. Ask your teacher what additional supplies (if any) will be available to you.

OBJECTIVES

- **Design** an experiment to solve a chemical problem.

- **Relate** observations of chemical properties to identify unknowns.

- **Infer** a conclusion from experimental data.

- **Apply** acid-base concepts.

MATERIALS

- 24-well microplate or 24 small test tubes

- labeled pipets containing solutions numbered 1–8

- toothpicks

  For other supplies, check with your teacher

Always wear safety goggles and a lab apron to protect your eyes and clothing. If you get a chemical in your eyes, immediately flush the chemical out at the eyewash station while calling to your teacher. Know the locations of the emergency lab shower and the eyewash station and the procedures for using them.

Do not touch any chemicals. If you get a chemical on your skin or clothing, wash the chemical off at the sink while calling to your teacher. Make sure you carefully read the labels and follow the precautions on all containers of chemicals that you use. If there are no precautions stated on the label, ask your teacher what precautions you should follow. Do not taste any chemicals or items used in the

Name _____ Class _____ Date _____

Is It an Acid or a Base? *continued*

laboratory. Never return leftovers to their original container; take only small amounts to avoid wasting supplies.

Never put broken glass in a regular waste container. Broken glass should be disposed of separately according to your teacher's instructions.

PREPARATION

1. Write the steps you will use to determine which solutions are acids and which solutions are bases.

   Student answers will vary, but the student should have a logical method of

   distinguishing which solutions are acid or base.

   _____

   _____

   _____

   _____

2. Ask your teacher to approve your plan and give you any additional supplies you will need.

3. Record your data in the data tables provided. In **Data Table 1,** record the numbers of the unknown solutions in the correct columns as you identify them. In **Data Table 2,** record the concentration of each solution as you test it, and then record the concentrations of HCl and NaOH present in the solution.

PROCEDURE

1. Carry out your plan for determining which solutions are acids and which are bases. As you perform your tests, avoid letting the tips of the storage pipets come into contact with other chemicals. Squeeze drops out of the pipets onto the 24-well plate and then use these drops for your tests. Record all observations in the space provided and then record your results in **Data Table 1.**

2. Write your procedure for determining the concentrations of the solutions. Ask your teacher to approve your plan, and request any additional supplies you will need.

   Student answers will vary, but the student should have a logical method,

   involving counting drops, of identifying the concentration of each solution.

   _____

   _____

   _____

   _____

Name _____ Class _____ Date _____

Is It an Acid or a Base? *continued*

3. Carry out your procedure for determining the concentrations of the solutions. Record all observations, and record your results in the second data table.

DISPOSAL

4. Clean all apparatus and your lab station. Return equipment to its proper place. Dispose of chemicals and solutions in the containers designated by your teacher. Do not pour any chemicals down the drain or in the trash unless your teacher directs you to do so. Wash your hands thoroughly before you leave the lab and after all work is finished.

OBSERVATIONS

Procedure 1 Student answers will vary, but should clearly show how solutions were identified as acid or base.

_____

_____

_____

_____

DATA TABLE 1	
Acids	**Bases**
2	1
3	4
6	5
8	7

Procedure 2 Student answers will vary, but should show the number of drops of each solution involved in a concentration determination.

_____

_____

_____

_____

Is It an Acid or a Base? *continued*

DATA TABLE 2		
Concentration	HCl	NaOH
0.8 M	3	4
0.4 M	6	5
0.2 M	8	7
0.1 M	2	1

Conclusions

1. **Analyzing Conclusions** List the numbers of the solutions and their concentrations.

   Students' answers depend on the numbers assigned to the unknown

   solutions.

   _____

   _____

   _____

   _____

   _____

   _____

   _____

   _____

   _____

Is It an Acid or a Base? *continued*

2. Analyzing Conclusions Describe the test results that led you to identify some solutions as acids and others as bases. Explain how you determined the concentrations of the unknown solutions.

Students' answers may vary. Students may have tried to measure the

volumes of the standard solutions needed to neutralize the unknown

solutions in order to determine their concentrations, or they may have

gauged concentration by comparing reaction rates.

_____

_____

_____

_____

EXTENSIONS

1. Evaluating Methods Compare your results with those of another lab group. Do you think that your teacher gave both groups the same set of solutions? (Is your solution 1 the same as their solution 1, and so on?) Explain your reasoning.

Students' answers will vary, but if the order of the acids and bases,

determined by one group and recorded in the second data table, is different

from that of the other group, students should infer that the two groups do

not have the same set of solutions.

_____

_____

Is It an Acid or a Base? *continued*

2. **Applying Conclusions** Imagine that you are helping to clean out the school's chemical storeroom. You find a spill coming from a large unlabeled reagent bottle filled with a clear liquid. What tests would you do to quickly determine if the substance is acidic or basic?

One possible method of figuring out the solutions' identities by using only

the solutions and the well plate would be to take several 8-drop samples of

each of the solutions and count how many (if any) drops of each of the other

solutions are required to turn the phenolphthalein in the sample red and

how many (if any) are required to neutralize it again. The most concentrated

base will be the solution that is able to turn the greatest number of solu-

tions red with the same number of drops. The most concentrated acid will be

the solution that requires the most drops of the most concentrated base to

neutralize. The other solutions should be easy to identify after the most

concentrated of each type are determined. (Note: Caution students to record

the results carefully because the plate would have to be cleaned off several

times to perform as many as 56 tests.)

Testing the pH of Rainwater

Do you have acid precipitation in your area?

MATERIALS

- rainwater
- distilled water
- 500 mL jars
- thin, transparent metric ruler (± 0.1 cm)
- pH test paper: narrow range, ± 0.2–0.3, or pH meter

Always wear safety goggles and a lab apron to protect your eyes and clothing. If you get a chemical in your eyes, immediately flush the chemical out at the eyewash station while calling to your teacher. Know the locations of the emergency lab shower and the eyewash station and the procedures for using them.

PROCEDURE

Record all of your results in a data table.

1. Each time it rains, set out five clean jars to collect the rainwater. If the rain continues for more than 24 hours, put out new containers at the end of each 24-hour period until the rain stops. (The same procedure can be used with snow if the snow is allowed to melt before measurements are taken. You may need to use larger containers if a heavy snowfall is expected.)

2. After the rain stops or at the end of each 24-hour period, use a thin, plastic ruler to measure the depth of the water to the nearest 0.1 cm. Using the pH paper, test the water to determine its pH to the nearest 0.2 to 0.3.

3. Record the following information:

   a. the date and time the collection started

   b. the date and time the collection ended

   c. the location where the collection was made (town and state)

   d. the amount of rainfall in centimeters

   e. the pH of the rainwater

4. Find the average pH of each collection that you have made for each rainfall, and record it in **Data Table 1.**

5. Collect samples on at least five different days. The more samples you collect, the more informative your data will be. Make up additional data sheets to record your data.

6. For comparison, determine the pH of pure water by testing five samples of distilled water with pH paper. Record your results in **Data Table 2,** and then calculate an average pH for distilled water.

Testing the pH of Rainwater *continued*

Jar	Start (date & time)	End (date & time)	Location	Rainfall (cm)	pH
					Data Table 1
1	*	*	*	*	*
2	*	*	*	*	*
3	*	*	*	*	*
4	*	*	*	*	*
5	*	*	*	*	*
					Avg. __*__

*Answers may vary.**

Sample	pH
	Data Table 2
1	*
2	*
3	*
4	*
5	*
	Avg. __*__

*Answers may vary.**

DISCUSSION

1. What is the pH of distilled water?

   7.0 ± 0.2; may be lower because of dissolved carbon dioxide

2. What is the pH of normal rainwater? How do you explain any differences between the pH readings?

   5.5–5.8 (see Table 3); Dissolved CO_2 and some pollutants, such as sulfur

   dioxide, lower pH, so the pH will vary slightly according to the amount of

   dissolved CO_2 or pollutants.

Testing the pH of Rainwater *continued*

3. What are the drawbacks of using a ruler to measure the depth of collected water? How could you increase the precision of your measurement?

Answers could include optical distortion, the fact that submerging the ruler

raises the water level, the formation of a meniscus when the ruler is too

close to the side of the jar, and the fact that the bottom of the jar may not

be flat. Use a calibrated rain gauge with markings on the glass to increase

precision.

_____

4. Does the amount of rainfall or the time of day the sample is taken have an effect on its pH? Try to explain any variability among samples.

Answers will vary depending on results.

_____

5. What conclusion can you draw from this investigation? Explain how your data support your conclusion.

Answers will vary. If the pH is consistently below 5.5, you may have acid

rain.

_____

_____

Name _____ Class _____ Date _____

How Much Calcium Carbonate Is in an Eggshell?

The calcium carbonate content of eggshells can be easily determined by means of an acid/base back-titration. In this experiment, a strong acid will react with calcium carbonate in eggshells. Then, the amount of unreacted acid will be determined by titration with a strong base.

OBJECTIVES

- **Determine** the amount of calcium carbonate present in an eggshell.

- **Relate** experimental titration measurements to a balanced chemical equation.

- **Infer** a conclusion from experimental data.

- **Apply** reaction stoichiometry concepts.

MATERIALS

- 1.00 M HCl
- 1.00 M NaOH
- 10 mL graduated cylinder
- 50 mL micro solution bottle or small Erlenmeyer flask
- 100 mL beaker
- balance
- desiccator (optional)
- distilled water

- drying oven
- eggshell
- forceps
- mortar and pestle
- phenolphthalein solution
- thin-stemmed pipets or medicine droppers, 3
- weighing paper

Always wear safety goggles and a lab apron to protect your eyes and clothing. If you get a chemical in your eyes, immediately flush the chemical out at the eyewash station while calling to your teacher. Know the locations of the emergency lab shower and the eyewash station and the procedures for using them.

Do not touch any chemicals. If you get a chemical on your skin or clothing, wash the chemical off at the sink while calling to your teacher. Make sure you carefully read the labels and follow the precautions on all containers of chemicals that you use. If there are no precautions stated on the label, ask your teacher what precautions you should follow. Do not taste any chemicals or items used in the laboratory. Never return leftovers to their original container; take only small amounts to avoid wasting supplies.

Acids and bases are corrosive. If an acid or base spills onto your skin or clothing, wash the area immediately with running water. Call your teacher in the event of an acid spill. Acid or base spills should be cleaned up promptly.

How Much Calcium Carbonate Is in an Eggshell? *continued*

PREPARATION

1. Wash an empty eggshell with distilled water and carefully peel all the membranes from its inside. Place *all* of the shell in a premassed beaker and dry the shell in the drying oven at 110°C for about 15 min.

2. Use **Data Table 1** and **Data Table 2** to record your data.

3. Put exactly 5.0 mL of water in the 10.0 mL graduated cylinder. Record this volume in **Data Table 1.** Label the first pipet *Acid.* To calibrate the pipet, fill it with water. **Do not use this pipet for the base solution.** Holding the pipet vertically, add 20 drops of water to the cylinder. Record the new volume of water in the graduated cylinder in **Data Table 1** under Trial 1.

4. Without emptying the graduated cylinder, add an additional 20 drops from the pipet. Record the new volume for Trial 2. Repeat this procedure once more for Trial 3.

5. Repeat Preparation steps **3** and **4** for the second pipet. Label this pipet *Base.* **Do not use this pipet for the acid solution.**

6. Make sure that the three trials produce data that are similar to one another. If one is greatly different from the others, perform Preparation steps **3–5** again.

7. Remove the eggshell and beaker from the oven. Cool them in a desiccator. Record the mass of the entire eggshell in **Data Table 2.** Place half of the shell into the clean mortar, and grind the shell into a very fine powder.

PROCEDURE

1. Measure the mass of a piece of weighing paper. Transfer about 0.1 g of ground eggshell to a piece of weighing paper, and measure the eggshell's mass as accurately as possible. Record the mass in **Data Table 2.** Place this eggshell sample into a clean, 50 mL micro solution bottle (or Erlenmeyer flask).

2. Fill the acid pipet with 1.00 M HCl acid solution, and then empty the pipet into an extra 100 mL beaker. Label the beaker "Waste." Fill the base pipet with the 1.00 M NaOH base solution, and then empty the pipet into the waste beaker.

3. Fill the acid pipet once more with 1.00 M HCl. Holding the acid pipet vertically, add exactly 150 drops of 1.00 M HCl to the bottle or flask that contains the eggshell. Swirl the flask gently for 3 to 4 min. Observe the reaction taking place. Wash down the sides of the flask with about 10 mL of distilled water. Using a third pipet, add two drops of phenolphthalein solution.

4. Fill the base pipet with the 1.00 M NaOH. Slowly add NaOH from the base pipet into the bottle or flask that contains the eggshell reaction mixture. Stop adding base when the mixture remains a faint pink color, even after it is swirled gently. **Be sure to add the base drop by drop, and be certain the drops end up in the reaction mixture and not on the walls of the bottle or flask.** Keep careful count of the number of drops used. Record in **Data Table 2** the number of drops of base used.

How Much Calcium Carbonate Is in an Eggshell? *continued*

DISPOSAL

5. Clean all equipment and your lab station. Dispose of chemicals and solutions as directed by your teacher. Wash your hands thoroughly before you leave the lab.

Data Table 1

Graduated Cylinder Readings (Pipet Calibration: Steps 3–5)

Trial	Initial—acid pipet (mL)	Final—acid pipet (mL)	Initial—base pipet (mL)	Final—base pipet (mL)
1	0.00	0.82	0.00	0.84
2	0.82	1.60	0.84	1.68
3	1.60	2.40	1.68	2.46

Total volume of drops—acid pipet ____60____

Average volume of each drop __0.040 mL__

Total volume of drops—base pipet ____60____

Average volume of each drop __0.041 mL__

Data Table 2

Data Table 2	
Mass of entire eggshell (g)	**mass will vary**
Mass of ground eggshell sample (g)	**0.11**
Number of drops of 1.00 M HCl added	**150 drops**
Volume of 1.00 M HCl added (mL)	**6.00**
Moles of HCl added	6.00×10^{-3}
Number of drops of 1.00 M NaOH added	**99 drops**
Volume of 1.00 M NaOH added (mL)	**4.06**
Moles of NaOH added	4.06×10^{-3}
Number of moles of HCl reacted with eggshell	1.94×10^{-3}
Number of moles of $CaCO_3$ reacted with HCl	9.70×10^{-4}
Mass of $CaCO_3$ in eggshell sample (g)	**0.0971**
% of $CaCO_3$ in eggshell sample	**88%**

How Much Calcium Carbonate Is in an Eggshell? *continued*

Analysis

1. **Organizing Ideas:** The calcium carbonate in the eggshell sample undergoes a double-displacement reaction with the HCl in step **3.** Write a balanced chemical equation for this reaction. (Hint: The gas observed was CO_2.)

   $CaCO_3(s) + 2HCl(aq) \rightarrow CO_2(g) + H_2O(l) + CaCl_2(aq)$

2. **Organizing Ideas:** Write the balanced chemical equation for the acid/base neutralization of the excess unreacted HCl with the NaOH.

   $HCl(aq) + NaOH(aq) \rightarrow NaCl(aq) + H_2O(l)$

3. **Organizing Data:** Calculate the volume of each drop in milliliters. Then convert the number of drops of HCl into volume in milliliters. Record this volume in **Data Table 2.** Repeat this step for the drops of NaOH.

   Sample calculation for acid pipet: 60 drops is equivalent to 2.40 mL;

   $\dfrac{2.40 \text{ mL}}{60 \text{ drops}} = 0.040 \text{ mL/drop}$

   Sample calculation for base pipet: 60 drops is equivalent to 2.46 mL;

   $\dfrac{2.46 \text{ mL}}{60 \text{ drops}} = 0.041 \text{ mL/drop}$

   150 drops of acid used × 0.040 mL/drop = 6.00 mL of acid used

   99 drops of base used × 0.041 mL/drop = 4.06 mL of base used

4. **Organizing Data:** Using the relationship between the molarity and volume of acid and the molarity and volume of base needed to neutralize it, calculate the volume of the HCl solution that was neutralized by the NaOH, and record it in **Data Table 2.** (Hint: This relationship was discussed in Section 2.)

   Students' answers will vary. The following is a sample calculation. Item 2 has

   a 1:1 ratio between acid and base in the neutralization reaction, so the

   following relation may be used:

   $mol_{acid} = (1.0 \text{ mol/L})(6.00 \text{ mL})(1 \text{ L/1000 mL}) = 6.00 \times 10^{-3} \text{ mol HCl}$

   $mol_{base} = (1.0 \text{ mol/L})(4.06 \text{ mL})(1 \text{ L/1000 mL}) = 4.06 \times 10^{-3} \text{ mol NaOH}$

How Much Calcium Carbonate Is in an Eggshell? *continued*

5. Analyzing Results: Calculate the volume and the number of moles of HCl that reacted with the $CaCO_3$ and record both in **Data Table 2.**

If the moles of HCl originally added is known and the moles of excess acid is

determined by the titration, then the difference will be the moles of HCl that

reacted with $CaCO_3$. The eggshell was dissolved in 6.00×10^{-3} moles of HCl.

The NaOH neutralized 4.06×10^{-3} moles of HCl. Therefore, the difference of

1.94×10^{-3} moles of HCl represents the amount of acid that reacted with

the eggshell.

_____

Conclusions

1. Organizing Data: Use the balanced equation for the reaction to calculate the number of moles of $CaCO_3$ that reacted with the HCl, and record this number in **Data Table 2.**

$$1.94 \times 10^{-3} \text{ mol HCl} \times \frac{1 \text{ mol CaCO}_3}{2 \text{ mol HCl}} = 9.70 \times 10^{-4} \text{ mol CaCO}_3$$

2. Organizing Data: Use the periodic table to calculate the molar mass of $CaCO_3$. Then, use the number of moles of $CaCO_3$ to calculate the mass of $CaCO_3$ in your eggshell sample. Record this mass in **Data Table 2.** Using the mass of $CaCO_3$, calculate the percentage of $CaCO_3$ in your eggshell and record it in **Data Table 2.**

$$9.70 \times 10^{-4} \text{ mol CaCO}_3 \times \frac{100.09 \text{ g}}{1 \text{ mol}} = 0.0971 \text{ g CaCO}_3$$

$$\frac{0.0971 \text{ g CaCO}_3}{0.11 \text{ g eggshell}} \times 100 = 88\% \text{ CaCO}_3$$

Skills Practice

Calorimetry and Hess's Law

Hess's law states that the overall enthalpy change in a reaction is equal to the sum of the enthalpy changes in the individual steps in the process. In this experiment, you will use a calorimeter to measure the energy released in three chemical reactions. From your experimental data, you will verify Hess's law.

OBJECTIVES

- **Demonstrate** proficiency in the use of calorimeters and related equipment.
- **Relate** temperature changes to enthalpy changes.
- **Determine** enthalpies of reaction for several reactions.
- **Demonstrate** that enthalpies of reactions can be additive.

MATERIALS

- 4 g NaOH pellets
- 50 mL 1.0 M HCl acid solution
- 50 mL 1.0 M NaOH solution
- 100 mL 0.50 M HCl solution
- 100 mL graduated cylinder
- balance
- distilled water

- forceps
- glass stirring rod
- gloves
- plastic-foam cups (or calorimeter)
- spatula
- thermometer
- watch glass

Always wear safety goggles and a lab apron to protect your eyes and clothing. If you get a chemical in your eyes, immediately flush the chemical out at the eyewash station while calling to your teacher. Know the locations of the emergency lab shower and the eyewash station and the procedures for using them.

Do not touch any chemicals. If you get a chemical on your skin or clothing, wash the chemical off at the sink while calling to your teacher. Make sure you carefully read the labels and follow the precautions on all containers of chemicals that you use. If there are no precautions stated on the label, ask your teacher what precautions you should follow. Do not taste any chemicals or items used in the laboratory. Never return leftovers to their original container; take only small amounts to avoid wasting supplies.

PREPARATION

1. Use the **Data Table** provided to record the total volumes of liquid, initial temperature, and final temperature of the three reactions you will carry out, as well as the mass of the empty watch glass and the watch glass plus NaOH pellets.

| Calorimetry and Hess's Law *continued*

2. Gently insert the thermometer into the plastic foam cup held upside down. **Thermometers break easily, so be careful with them, and do not use them to stir a solution.**

PROCEDURE

Reaction 1: Dissolving NaOH

1. Pour 100 mL of distilled water into your calorimeter. Record the water temperature to the nearest 0.1°C.

2. Weigh a clean and dry watch glass to the nearest 0.01 g. Wearing gloves and using forceps, place about 2 g of NaOH pellets on the watch glass. Measure and record the mass of the watch glass and the pellets to the nearest 0.01 g. **It is important that this step be done quickly: NaOH absorbs moisture from the air.**

3. Immediately place the NaOH pellets in the calorimeter cup, and gently stir the solution with a stirring rod. **Do not stir with a thermometer.** Place the lid on the calorimeter. Watch the thermometer, and record the highest temperature in the **Data Table.**

4. Be sure to clean all equipment and rinse it with distilled water before continuing.

Reaction 2: NaOH and HCl in Solution

5. Pour 50 mL of 1.0 M HCl into your calorimeter. Record the temperature of the HCl solution to the nearest 0.1°C.

6. Pour 50 mL of 1.0 M NaOH into a graduated cylinder. **For this step only, rinse the thermometer, and measure the temperature of the NaOH solution in the graduated cylinder to the nearest 0.1°C. Record the temperature, and then replace the thermometer in the calorimeter.**

7. Pour the NaOH solution into the calorimeter cup, and stir gently. Place the lid on the calorimeter. Watch the thermometer and record the highest temperature.

8. Pour the solution in the container designated by your teacher. Clean and rinse all equipment before continuing with the procedure.

Reaction 3: Solid NaOH and HCl in Solution

9. Pour 100 mL of 0.50 M HCl into your calorimeter. Record the temperature of the HCl solution to the nearest 0.1°C.

10. Measure the mass of a clean and dry watch glass, and record the mass. Wear gloves, and using forceps, obtain approximately 2 g of NaOH pellets. Place them on the watch glass, and record the total mass. **As in step 2, it is important that this step be done quickly.**

Calorimetry and Hess's Law *continued*

11. Immediately place the NaOH pellets in the calorimeter, and gently stir the solution. Place the lid on the calorimeter. Watch the thermometer, and record the highest temperature. When finished with this reaction, pour the solution into the container designated by your teacher for disposal of mostly neutral solutions.

DISPOSAL

12. Check with your teacher for the proper disposal procedures. Always wash your hands thoroughly after cleaning up the lab area and equipment.

Data Table			
Reaction	**1**	**2**	**3**
Volume of liquid(s), mL	100.0	100.0	100.0
Initial temperature, °C	21.5	22.0	22.0
Highest temperature, °C	26.5	28.1	33.0
Mass of empty watch glass, g	30.15		30.15
Mass of watch glass + NaOH, g	32.15		32.15

Analysis

1. Organizing Ideas Write a balanced chemical equation for each of the three reactions that you performed. (Hint: Be sure to include the physical states of matter for all substances.)

$$NaOH(s) \longrightarrow NaOH(aq)$$

$$NaOH(aq) + HCl(aq) \longrightarrow H_2O(l) + NaCl(aq)$$

$$NaOH(s) + HCl(aq) \longrightarrow H_2O(l) + NaCl(aq)$$

2. Organizing Ideas Write the equation for the total reaction by adding two of the equations from item **1** and then canceling out substances that appear in the same form on both sides of the new equation.

equation 1 + equation 2 = equation 3

Calorimetry and Hess's Law *continued*

3. **Organizing Data** Calculate the change in temperature for each of the reactions.

$\Delta t_1 = 26.5°C - 21.5°C = 5.0°C$

$\Delta t_2 = 28.1°C - 22.0°C = 6.1°C$

$\Delta t_3 = 33.0°C - 22.0°C = 11.0°C$

4. **Organizing Data** Assuming that the density of the water and the solutions is 1.00 g/mL, calculate the mass of liquid present for each of the reactions.

$m = 100.0 \text{ mL } H_2O \times \dfrac{1.00 \text{ g}}{1 \text{ mL } H_2O} = 100.0 \text{ g } H_2O \text{ for all three reactions}$

5. **Analyzing Results** Using the calorimeter equation, calculate the energy as heat released by each reaction. (Hint: Use the specific heat of water in your calculations.)

$$c_p, H_2O = 4.184 \text{ J/g} \bullet °C$$

$$\text{heat} = m \times \Delta t \times c_p, H_2O$$

Energy for reaction 1:

$100.0 \text{ g } H_2O \times 5.0°C \times \dfrac{4.184 \text{ J}}{1 \text{ g} \cdot °C} = 2100 \text{ J} = 2.1 \text{ kJ}$

Energy for reaction 2:

$100.0 \text{ g } H_2O \times 6.1°C \times \dfrac{4.184 \text{ J}}{1 \text{ g} \cdot °C} = 2500 \text{ J} = 2.5 \text{ kJ}$

Energy for reaction 3:

$100.0 \text{ g } H_2O \times 11.0°C \times \dfrac{4.184 \text{ J}}{1 \text{ g} \cdot °C} = 4600 \text{ J} = 4.6 \text{ kJ}$

6. **Organizing Data** Calculate the moles of NaOH used in each of the reactions.

Moles NaOH for reaction 1:

$2.00 \text{ g} \times \dfrac{1 \text{ mol NaOH}}{40.00 \text{ g}} = 5.00 \times 10^{-2} \text{ mol}$

Moles NaOH for reaction 2:

$50 \text{ mL} \times \dfrac{1 \text{ L}}{1000 \text{ mL}} \times \dfrac{1.00 \text{ mol NaOH}}{1 \text{ L}} = 5.00 \times 10^{-2} \text{ mol}$

Moles NaOH for reaction 3:

$2.01 \text{ g} \times \dfrac{1 \text{ mol NaOH}}{40.00 \text{ g}} = 5.02 \times 10^{-2} \text{ mol}$

Calorimetry and Hess's Law *continued*

7. **Analyzing Results** Calculate the ΔH value in kJ/mol of NaOH for each of the three reactions.

$$\Delta H_1 = \frac{-2.1 \text{ kJ}}{5.00 \times 10^{-2} \text{ mol NaOH}} = -42 \text{ kJ/mol NaOH}$$

$$\Delta H_2 = \frac{-2.5 \text{ kJ}}{5.00 \times 10^{-2} \text{ mol NaOH}} = -50 \text{ kJ/mol NaOH}$$

$$\Delta H_3 = \frac{-4.6 \text{ kJ}}{5.02 \times 10^{-2} \text{ mol NaOH}} = -92 \text{ kJ/mol NaOH}$$

8. **Organizing Ideas** Using your answer to Analysis item **2** and your knowledge of Hess's law, explain how the enthalpies for the three reactions should be mathematically related.

The sum of enthalpies for the first two reactions should equal the enthalpy

for the third reaction.

Name _____ Class _____ Date _____

Factors Influencing Reaction Rate

How do the type of reactants, surface area of reactants, concentration of reactants, and catalysts affect the rates of chemical reactions?

MATERIALS

- Bunsen burner
- paper ash
- copper foil strip
- graduated cylinder, 10 mL
- magnesium ribbon
- matches
- paper clip

- sandpaper
- steel wool
- 2 sugar cubes
- white vinegar
- zinc strip
- 6 test tubes, 16 × 150 mm
- tongs

Always wear safety goggles and a lab apron to protect your eyes and clothing. If you get a chemical in your eyes, immediately flush the chemical out at the eyewash station while calling to your teacher. Know the locations of the emergency lab shower and the eyewash station and the procedures for using them.

PROCEDURE

Remove all combustible material from the work area.

1. Add 10 mL of vinegar to each of three test tubes. To one test tube, add a 3 cm piece of magnesium ribbon; to a second, add a 3 cm zinc strip; and to a third, add a 3 cm copper strip. (All metals should be the same width.) If necessary, polish the metals with sandpaper until they are shiny. Record your results in **Data Table 1.**

2. Using tongs, hold a paper clip in the hottest part of the burner flame for 30 s. Repeat with a ball of steel wool 2 cm in diameter. Record your results in **Data Table 2.**

3. To one test tube, add 10 mL of vinegar; to a second, add 5 mL of vinegar plus 5 mL of water; and to a third, add 2.5 mL of vinegar plus 7.5 mL of water. To each of the three test tubes, add a 3 cm piece of magnesium ribbon. Record your results in **Data Table 3.**

4. Using tongs, hold a sugar cube and try to ignite it with a match. Then try to ignite it in a burner flame. Rub paper ash on a second cube, and try to ignite it with a match. Record your results in **Data Table 4.**

DISPOSAL

5. Combine all liquids and pour them down the drain. Save all metal strips for reuse, but if they are too corroded, put them in the trash. Put all other solids in the trash.

Name _____ Class _____ Date _____

Factors Influencing Reaction Rate *continued*

Data Table 1		
10 mL vinegar + Mg ribbon	10 mL vinegar + Zn strip	10 mL vinegar + Cu strip
vigorous bubbling	bubbling	no bubbling

Data Table 2	
Paper clip in flame	Steel wool in flame
discolors	burns quickly

Data Table 3		
10 mL vinegar + Mg ribbon	5 mL vinegar + 5 mL H_2O + Mg ribbon	2.5 mL vinegar + 7.5 mL water + Mg ribbon
vigorous bubbling	bubbling	less bubbling

Data Table 4		
Sugar cube with match	Sugar cube with burner	Sugar cube + paper ash with match
some melting and turns dark brown	turns black, some melting	turns black very quickly

DISCUSSION

1. What are the rate-influencing factors in each step of the procedure?

1: the nature of the materials; 2: surface area of the metal; 3: concentration

of the reactant; 4: the presence of a catalyst

2. What were the results from each step of the procedure? How do you interpret each result?

1: Mg reacts faster than Zn, which reacts faster than Cu; nature of reactants

affects reaction rate

2: steel wool reacts faster with oxygen than a paper clip does; increasing

surface area increases rate of reaction

3: Mg reacts faster with more concentrated vinegar; increased concentration

increases reaction rate

4: sugar reacts faster when covered with ash than without; catalysts increase

rate of reaction

Name _____ Class _____ Date _____

(Microscale)

Rate of a Chemical Reaction

In this experiment, you will determine the rate of the reaction whose net equation is written as follows:

$$3Na_2S_2O_5(aq) + 2KIO_3(aq) + 3H_2O(l) \xrightarrow{H^+} 2KI(aq) + 6NaHSO_4(aq)$$

One way to study the rate of this reaction is to observe how fast $Na_2S_2O_5$ is used up. After all the $Na_2S_2O_5$ solution has reacted, the concentration of iodine, I_2, an intermediate in the reaction, increases. A starch indicator solution, added to the reaction mixture, will change from clear to a blue-black color in the presence of I_2.

In the procedure, the concentrations of the reactants are given in terms of drops of solution A and drops of solution B. Solution A contains $Na_2S_2O_5$, the starch indicator solution, and dilute sulfuric acid to supply the hydrogen ions needed to catalyze the reaction. Solution B contains KIO_3. You will run the reaction with several different concentrations of the reactants and record the time it takes for the blue-black color to appear.

OBJECTIVES

- **Prepare** and **observe** several different reaction mixtures.

- **Demonstrate** proficiency in measuring reaction rates.

- **Relate** experimental results to a rate law that can be used to predict the results of various combinations of reactants.

MATERIALS

- 8-well microscale reaction strips, 2

- distilled or deionized water

- fine-tipped dropper bulbs or small microtip pipets, 3

- solution A

- solution B

- stopwatch or clock with second hand

Always wear safety goggles and a lab apron to protect your eyes and clothing. If you get a chemical in your eyes, immediately flush the chemical out at the eyewash station while calling to your teacher. Know the locations of the emergency lab shower and the eyewash station and the procedures for using them.

PREPARATION

1. Use the **Data Table** provided to record your data.

2. Obtain three dropper bulbs or small microtip pipets, and label them *A*, *B*, and *H_2O*.

3. Fill the bulb of pipet A with solution A, the bulb of pipet B with solution B, and the bulb of pipet for H_2O with distilled water.

PROCEDURE

1. Using the first 8-well strip, place five drops of solution A into each of the first five wells. Record the number of drops in the appropriate places in the **Data Table. For the best results, try to make all drops about the same size.**

2. In the second 8-well reaction strip, place one drop of solution B in the first well, two drops in the second well, three drops in the third well, four drops in the fourth well, and five drops in the fifth well. Record the number of drops in the **Data Table.**

3. In the second 8-well strip that contains drops of solution B, add four drops of water to the first well, three drops to the second well, two drops to the third well, and one drop to the fourth well. Do not add any water to the fifth well.

4. Carefully invert the second strip. The surface tension should keep the solutions from falling out of the wells. Place the strip well-to-well on top of the first strip.

5. Hold the strips tightly together and record the exact time, or set the stop-watch, as you shake the strips once. This procedure should mix the upper solutions with each of the corresponding lower ones.

6. Observe the lower wells. Note the sequence in which the solutions react, and record the number of seconds it takes for each solution to turn a blue-black color.

DISPOSAL

7. Dispose of the solutions in the container designated by your teacher. Wash your hands thoroughly after cleaning up the area and equipment.

Data Table					
Well	**1**	**2**	**3**	**4**	**5**
Time rxn. began	00:00	00:00	00:00	00:00	00:00
Time rxn. stopped	02:23	01:07	00:43	00:35	00:26
Drops of A	5	5	5	5	5
Drops of B	1	2	3	4	5
Drops of H_2O	4	3	2	1	0

Name _____ Class _____ Date _____

| Rate of a Chemical Reaction *continued*

Analysis

1. **Organizing Data:** Calculate the time elapsed for the complete reaction of each combination of solutions A and B.

   Students will need to convert from minutes and seconds to seconds.

   Well 1: 143 s   **Well 4: 35 s**

   Well 2: 67 s   **Well 5: 26 s**

   Well 3: 43 s

2. **Evaluating Data:** Make a graph of your results. Label the x-axis "Number of drops of solution B." Label the y-axis "Time elapsed." Make a similar graph for drops of solution B versus rate (1/time elapsed).

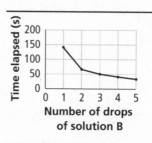

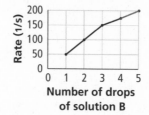

3. **Analyzing Information:** Which mixture reacted the fastest? Which mixture reacted the slowest?

   The most concentrated solution (5 drops of solution B) had the fastest

   reaction time. The least concentrated solution (1 drop of solution B and

   4 drops of water) had the slowest reaction time.

4. Evaluating Methods: Why was it important to add the drops of water to the wells that contained fewer than five drops of solution B? (Hint: Figure out the total number of drops in each of the reaction wells.)

<u>The total number of drops for each well should be the same to make valid</u>

<u>comparisons of different concentrations.</u>

_____

_____

Conclusions

1. Evaluating Conclusions: Which of the following variables that can affect the rate of a reaction is tested in this experiment: temperature, catalyst, concentration, surface area, or nature of reactants? Explain your answer.

<u>The effect of reactant concentration (solution B) on reaction rate is being</u>

<u>measured.</u>

_____

2. Applying Ideas: Use your data and graphs to determine the relationship between the concentration of solution B and the rate of the reaction. Describe this relationship in terms of a rate law.

<u>The rate of the reaction is directly proportional to the concentration of</u>

<u>solution B.</u>

EXTENSIONS

1. Predicting Outcomes: What combination of drops of solutions A and B would you use if you wanted the reaction to last exactly 2.5 min?

<u>Students' answers will vary, depending on the conditions in your lab. For the</u>

<u>sample data, the reaction would require a mixture of 4 drops of A, 25 drops of</u>

<u>B, and 21 drops of H_2O. Students may need to use a different reaction vessel if</u>

<u>this amount of solution is too much for the microwell strips. Be sure answers</u>

<u>are safe, and include carefully planned procedures.</u>

_____

Name _____ Class _____ Date _____

Measuring K_a for Acetic Acid

The acid dissociation constant, K_a, is a measure of the strength of an acid. Strong acids are completely ionized in water. Because weak acids are only partly ionized, they have a characteristic K_a value. Properties that depend on the ability of a substance to ionize, such as conductivity and colligative properties, can be used to measure K_a. In this experiment, you will compare the conductivity of a 1.0 M solution of acetic acid, CH_3COOH, a weak acid, with the conductivities of solutions of varying concentrations of hydrochloric acid, HCl, a strong acid. From the comparisons you make, you will be able to estimate the concentration of hydronium ions in the acetic acid solution and calculate its K_a.

OBJECTIVES

- **Compare** the conductivities of solutions of known and unknown hydronium ion concentrations.

- **Relate** conductivity to the concentration of ions in solution.

- **Explain** the validity of the procedure on the basis of the definitions of strong and weak acids.

- **Compute** the numerical value of K_a for acetic acid.

MATERIALS

- 1.0 M acetic acid, CH_3COOH
- 1.0 M hydrochloric acid, HCl
- 24-well plate
- distilled or deionized water

- LED conductivity testers
- paper towels
- thin-stemmed pipets

Always wear safety goggles and a lab apron to protect your eyes and clothing. If you get a chemical in your eyes, immediately flush the chemical out at the eyewash station while calling to your teacher. Know the locations of the emergency lab shower and the eyewash station and the procedures for using them.

Do not touch any chemicals. If you get a chemical on your skin or clothing, wash the chemical off at the sink while calling to your teacher. Make sure you carefully read the labels and follow the precautions on all containers of chemicals that you use. If there are no precautions stated on the label, ask your teacher what precautions you should follow. Do not taste any chemicals or items used in the laboratory. Never return leftovers to their original container; take only small amounts to avoid wasting supplies.

Acids and bases are corrosive. If an acid or base spills onto your skin or clothing, wash the area immediately with running water. Call your teacher in the event of an acid spill. Acid or base spills should be cleaned up promptly.

PROCEDURE

1. Obtain samples of 1.0 M HCl solution and 1.0 M CH₃COOH solution.

2. Place 20 drops of HCl in one well of a 24-well plate. Place 20 drops of CH₃COOH in an adjacent well. Label the location of each sample.

3. Test the HCl and CH₃COOH with the conductivity tester. Note the relative intensity of the tester light for each solution. After testing, rinse the tester probes with distilled water. Remove any excess moisture with a paper towel.

4. Place 18 drops of distilled water in each of six wells in your 24-well plate. Add 2 drops of 1.0 M HCl to the first well to make a total of 20 drops of solution. Mix the contents of this well thoroughly by picking the contents up in a pipet and returning them to the well.

5. Repeat this procedure by taking 2 drops of the previous dilution and placing it in the next well containing 18 drops of water. Return any unused solution in the pipet to the well from which it was taken. Mix the new solution with a new pipet. (You now have 1.0 M HCl in the well from Procedure step **2**, 0.10 M HCl in the first dilution well, and 0.010 M HCl in the second dilution.)

6. Continue diluting in this manner until you have six successive dilutions. The $[H_3O^+]$ should now range from 1.0 M to 1.0×10^{-6} M. Write the concentrations in the first column of the **Data Table** provided.

7. Using the conductivity tester, test the cells containing HCl in order from most concentrated to least concentrated. Note the brightness of the tester bulb, and compare it with the brightness of the bulb when it was placed in the acetic acid solution. (Retest the acetic acid well any time for comparison.) After each test, rinse the tester probes with distilled water, and use a paper towel to remove any excess moisture. When the brightness produced by one of the HCl solutions is about the same as that produced by the acetic acid, you can infer that the two solutions have about the same hydronium ion concentration and that the pH of the HCl solution is equal to the pH of the acetic acid. If the glow from the bulb is too faint to see, turn off the lights or build a light shield around your conductivity tester bulb.

8. Record the results of your observations by noting which HCl concentration causes the intensity of the bulb to most closely match that of the bulb when it is in acetic acid. (Hint: If the conductivity of no single HCl concentration matches that of the acetic acid, then estimate the value between the two concentrations that match the best.)

DISPOSAL

9. Clean your lab station. Clean all equipment, and return it to its proper place. Dispose of chemicals and solutions in containers designated by your teacher. Do not pour any chemicals down the drain or throw anything in the trash unless your teacher directs you to do so. Wash your hands thoroughly after all work is finished and before you leave the lab.

| Measuring K_a for Acetic Acid *continued*

DATA TABLE	
HCl concentration	**Observations and comparisons**
1.0 M	Bright light
0.10 M	Dimmer light
0.010 M	Dimmer light; nearly same brightness as 1.0 M CH_3OOH
1.0×10^{-3} M	Dimmer light; nearly same brightness as 1.0 M CH_3OOH
1.0×10^{-4} M	Dimmer light
1.0×10^{-5} M	Light barely visible
1.0×10^{-6} M	Light barely visible

Analysis

1. **Resolving Discrepancies** How did the conductivity of the 1.0 M HCl solution compare with that of the 1.0 M CH_3COOH solution? Why do you think this was so?

   The conductivity of the 1.0 M CH_3COOH solution was much less than that of

   the 1.0 M HCl solution. Hydrochloric acid is a strong acid and ionizes

   completely in water, but acetic acid is a weak acid and ionizes only partly in

   water. The greater number of ions in the HCl solution accounts for its

   greater conductivity.

2. **Organizing Data** What is the H_3O^+ concentration of the HCl solution that most closely matched the conductivity of the acetic acid?

   Students should find that either the 10^{-2} M or the 10^{-3} M HCl solution is

   the best match for the acetic acid solution. Some students will estimate that

   the $[H_3O^+]$ falls between these values.

Measuring K_a for Acetic Acid *continued*

3. Drawing Conclusions What was the H_3O^+ concentration of the 1.0 M CH_3COOH solution? Why?

The estimates for $[H_3O^+]$ should range between 10^{-2} M and 10^{-3} M because

these values match the values of the respective HCl solutions with the same

conductivity and thus have the same number of ions and the same $[H_3O^+]$ as

the HCl solutions.

Conclusions

1. Applying Models The acid ionization expression for CH_3COOH is the following:

$$K_a = \frac{[H_3O^+]\,[CH_3COO^-]}{[CH_3COOH]}$$

Use your answer to Analysis item **3** to calculate K_a for the acetic acid solution.

Student values for K_a will vary depending on student estimates of $[H_3O^+]$.

Those who estimate it as being close to 10^{-2} M should calculate a K_a value

of 10^{-4}. Those who estimate $[H_3O^+]$ as closer to 10^{-3} M should calculate a

value of 10^{-6}.

$$K_a = \frac{[H_3O^+]\,[CH_3COO^-]}{[CH_3COOH]} = \frac{(10^{-3})\,(10^{-3})}{(1.0 - 10^{-3})} = \frac{(10^{-3})^2}{1.0} = 10^{-6}$$

2. Applying Models Explain how it is possible for solutions of HCl and CH_3COOH to show the same conductivity but have different concentrations.

The conductivity of a solution is a function of the number of ions in the

solution. Both HCl and CH_3COOH ionize to form an anion and the hydronium

ion. When the conductivities of two solutions match, they each contain the

same total number of ions and the same number of hydronium ions even

though their concentrations differ. This occurs because HCl is a strong acid

and therefore ionizes completely. CH_3COOH is a weak acid and does not

ionize completely.

Measuring K_a for Acetic Acid *continued*

EXTENSIONS

1. **Evaluating Methods** Compare the K_a value that you calculated with the value found on page 606 of your text. Calculate the percent error for this experiment.

   Students' answers will vary.

2. **Predicting Outcomes** Lactic acid ($HOOCCHOHCH_3$) has a K_a of 1.4×10^{-4}. Predict whether a solution of lactic acid would cause the conductivity tester to glow brighter or dimmer than a solution of acetic acid with the same concentration. How noticeable would the difference be?

   Lactic acid has a larger K_a than acetic acid, so it will produce a more intense

   light in a conductivity tester. However, the glow produced by lactic acid

   should only be about as bright as that of the 0.01 M HCl solution because

   $[H_3O^+]$ in a 1.0 M solution of lactic acid is approximately 0.01. Because the

   conductivity of the acetic acid solution was between that of the 0.01 M HCl

   and that of the 0.001 M HCl, the difference will not be very great.

Quick Lab

Redox Reactions

MATERIALS

- aluminum foil
- beaker, 250 mL
- 1 M copper(II) chloride solution, $CuCl_2$
- 3% hydrogen peroxide
- manganese dioxide
- metric ruler
- scissors
- test-tube clamp
- test tube, 16 × 150 mm
- wooden splint

Always wear safety goggles and a lab apron to protect your eyes and clothing. If you get a chemical in your eyes, immediately flush the chemical out at the eyewash station while calling to your teacher. Know the locations of the emergency lab shower and the eyewash station and the procedures for using them.

Do not touch any chemicals. If you get a chemical on your skin or clothing, wash the chemical off at the sink while calling to your teacher. Make sure you carefully read the labels and follow the precautions on all containers of chemicals that you use. If there are no precautions stated on the label, ask your teacher what precautions you should follow. Do not taste any chemicals or items used in the laboratory. Never return leftovers to their original container; take only small amounts to avoid wasting supplies.

PROCEDURE

Record all of your results in the **Data Table.**

1. Put 10 mL of hydrogen peroxide in a test tube, and add a small amount of manganese dioxide (equal to the size of about half a pea). What is the result?

2. Insert a glowing wooden splint into the test tube. What is the result? If oxygen is produced, a glowing wooden splint inserted into the test tube will glow brighter.

3. Fill the 250 mL beaker halfway with the copper(II) chloride solution.

4. Cut foil into 2 cm × 12 cm strips.

5. Add the aluminum strips to the copper(II) chloride solution. Use a glass rod to stir the mixture, and observe for 12 to 15 minutes. What is the result?

DISPOSAL

6. Check with your teacher for the proper disposal procedures. Always wash your hands thoroughly after cleaning up the lab area and equipment.

Redox Reactions *continued*

Data Table	
Reactants	**Result**
H_2O_2 + MgO	**vigorous bubbling occurs**
H_2O_2 + MgO with glowing splint	**the splint ignites**
aluminum foil + copper(II) chloride	**the edges of the foil turn red-brown in color; the solution becomes less blue in color**

DISCUSSION

1. Write balanced equations showing what happened in each of the reactions.

$2H_2O_2 \rightarrow O_2 + 2H_2O$; $3CuCl_2 + 2Al \rightarrow 2AlCl_3 + 3Cu$

_____

2. Write a conclusion for the two experiments.

Hydrogen peroxide was oxidized to oxygen and reduced to water. Evidence

for the formation of O_2 was seen when a glowing splint burst into flames

when placed in the reaction test tube. Copper(II) ions were reduced to

copper metal, and aluminum metal was oxidized to Al(III) ions.

_____

Name _____ Class _____ Date _____

(Microscale)

Reduction of Manganese in Permanganate Ion

In Chapter 15, you studied acid-base titrations in which an unknown amount of acid is titrated with a carefully measured amount of base. In this procedure, a similar approach called a *redox titration* is used. In a redox titration, the reducing agent, Fe^{2+}, is oxidized to Fe^{3+} by the oxidizing agent, MnO_4^-. When this process occurs, the Mn in MnO_4^- changes from a +7 to a +2 oxidation state and has a noticeably different color. You can use this color change to signify a redox reaction "end point." When the reaction is complete, any excess MnO_4^- added to the reaction mixture will give the solution a pink or purple color. The volume data from the titration, the known molarity of the $KMnO_4$ solution, and the mole ratio from the balanced redox equation will give you the information you need to calculate the molarity of the $FeSO_4$ solution.

OBJECTIVES

- **Demonstrate** proficiency in performing redox titrations and recognizing end points of a redox reaction.

- **Write** a balanced oxidation-reduction equation for a redox reaction.

- **Determine** the concentration of a solution by using stoichiometry and volume data from a titration.

MATERIALS

- 0.0200 M $KMnO_4$
- 1.0 M H_2SO_4
- 100 mL graduated cylinder
- 125 mL Erlenmeyer flasks, 4
- 250 mL beakers, 2
- 400 mL beaker

- burets, 2
- distilled water
- double buret clamp
- $FeSO_4$ solution
- ring stand
- wash bottle

Always wear safety goggles and a lab apron to protect your eyes and clothing. If you get a chemical in your eyes, immediately flush the chemical out at the eyewash station while calling to your teacher. Know the locations of the emergency lab shower and the eyewash station and the procedures for using them.

Do not touch any chemicals. If you get a chemical on your skin or clothing, wash the chemical off at the sink while calling to your teacher. Make sure you carefully read the labels and follow the precautions on all containers of chemicals that you use. If there are no precautions stated on the label, ask your teacher what precautions you should follow. Do not taste any chemicals or items used in the laboratory. Never return leftovers to their original container; take only small amounts to avoid wasting supplies.

Reduction of Manganese in Permanganate Ion *continued*

 Acids and bases are corrosive. If an acid or base spills onto your skin or clothing, wash the area immediately with running water. Call your teacher in the event of an acid spill. Acid or base spills should be cleaned up promptly.

Never put broken glass in a regular waste container. Broken glass should be disposed of separately according to your teacher's instructions.

PREPARATION

1. Use the data table provided to record your data.

2. Clean two 50 mL burets with a buret brush and distilled water. Rinse each buret at least three times with distilled water to remove contaminants.

3. Label one 250 mL beaker *0.0200 M KMnO₄* and the other *FeSO₄*. Label three of the flasks *1*, *2*, and *3*. Label the 400 mL beaker *Waste*. Label one buret *KMnO₄* and the other *FeSO₄*.

4. Measure approximately 75 mL of 0.0200 M KMnO₄, and pour it into the appropriately labeled beaker. Obtain approximately 75 mL of FeSO₄ solution, and pour it into the appropriately labeled beaker.

5. Rinse one buret three times with a few milliliters of 0.0200 M KMnO₄ from the appropriately labeled beaker. Collect these rinses in the waste beaker. Rinse the other buret three times with small amounts of FeSO₄ solution from the appropriately labeled beaker. Collect these rinses in the waste beaker.

6. Set up the burets as instructed by your teacher. Fill one buret with approximately 50 mL of 0.0200 M KMnO₄ from the beaker, and fill the other buret with approximately 50 mL of the FeSO₄ solution from the other beaker.

7. With the waste beaker underneath its tip, open the KMnO₄ buret long enough to be sure the buret tip is filled. Repeat the process for the FeSO₄ buret.

8. Add 50 mL of distilled water to one of the 125 mL Erlenmeyer flasks, and add one drop of 0.0200 M KMnO₄ to the flask. Set this mixture aside to use as a color standard. It can be compared with the titration mixture to determine the end point.

PROCEDURE

1. Record in the **Data Table** the initial buret readings for both solutions. Add 10 mL of the hydrated iron(II) sulfate solution, $FeSO_4 \cdot 7H_2O$, to the flask labeled *1*. Add 5 mL of 1 M H_2SO_4 to the FeSO₄ solution in this flask. The acid will help keep the Fe^{2+} ions in the reduced state, which will allow you time to titrate.

2. Slowly add KMnO₄ from the buret to the FeSO₄ in the flask while swirling the flask. When the color of the solution matches the color standard you prepared in Preparation step **8,** record in the **Data Table** the final readings of the burets.

3. Empty the titration flask into the waste beaker. Repeat the titration procedure in steps **1** and **2** with the flasks labeled *2* and *3*.

▌Reduction of Manganese in Permanganate Ion *continued*

DISPOSAL

4. Dispose of the contents of the waste beaker in the container designated by your teacher. Also, pour the color-standard flask into this container. Wash your hands thoroughly after cleaning up the area and equipment.

Data Table				
Trial	Initial KMnO$_4$ volume (mL)	Final KMnO$_4$ volume (mL)	Initial FeSO$_4$ volume (mL)	Final FeSO$_4$ volume (mL)
1	50.0	35.0	50.0	40.0
2	35.0	20.5	40.0	30.0
3	20.5	5.0	30.0	20.0

Analysis

1. **Organizing Ideas:** Write the balanced equation for the redox reaction of FeSO$_4$ and KMnO$_4$.

$$\text{MnO}_4^-(aq) + 8\text{H}^+(aq) + 5\text{Fe}^{2+}(aq) \rightarrow 5\text{Fe}^{3+}(aq) + \text{Mn}^{2+}(aq) + 4\text{H}_2\text{O}(l)$$

2. **Evaluating Data:** Calculate the number of moles of MnO$_4^-$ reduced in each trial.

Trial 1:

$$15.0 \text{ mL KMnO}_4 \times \frac{1 \text{ L}}{1000 \text{ mL}} \times \frac{0.0200 \text{ mol KMnO}_4}{1 \text{ L}} = 3.00 \times 10^{-4} \text{ mol KMnO}_4$$

Trial 2:

$$14.5 \text{ mL KMnO}_4 \times \frac{1 \text{ L}}{1000 \text{ mL}} \times \frac{0.0200 \text{ mol KMnO}_4}{1 \text{ L}} = 2.90 \times 10^{-4} \text{ mol KMnO}_4$$

Trial 3:

$$15.5 \text{ mL KMnO}_4 \times \frac{1 \text{ L}}{1000 \text{ mL}} \times \frac{0.0200 \text{ mol KMnO}_4}{1 \text{ L}} = 3.10 \times 10^{-4} \text{ mol KMnO}_4$$

Reduction of Manganese in Permanganate Ion *continued*

3. Analyzing Information: Calculate the number of moles of Fe^{2+} oxidized in each trial.

The ratio of Fe^{2+} to MnO_4^- is 5:1.

Trial 1:

$$3.0 \times 10^{-4} \text{ mol } MnO_4^- \times \frac{5 \text{ mol } Fe^{2+}}{1 \text{ mol } MnO_4^-} = 1.5 \times 10^{-3} \text{ mol } Fe^{2+}$$

Trial 2:

$$2.9 \times 10^{-4} \text{ mol } MnO_4^- \times \frac{5 \text{ mol } Fe^{2+}}{1 \text{ mol } MnO_4^-} = 1.4 \times 10^{-3} \text{ mol } Fe^{2+}$$

Trial 3:

$$3.1 \times 10^{-4} \text{ mol } MnO_4^- \times \frac{5 \text{ mol } Fe^{2+}}{1 \text{ mol } MnO_4^-} = 1.6 \times 10^{-3} \text{ mol } Fe^{2+}$$

4. Applying Conclusions: Calculate the average concentration (molarity) of the iron(II) sulfate solution.

Trial 1:

$$\frac{1.5 \times 10^{-3} \text{ mol}}{0.01L} = 0.15 \text{ M}$$

Trial 2:

$$\frac{1.4 \times 10^{-3} \text{ mol}}{0.01L} = 0.14 \text{ M}$$

Trial 3:

$$\frac{1.6 \times 10^{-3} \text{ mol}}{0.01L} = 0.16 \text{ M}$$

Average molarity

$$\frac{0.15M + 0.14M + 0.16M}{3} = 0.15 \text{ M}$$

EXTENSIONS

1. Designing Experiments: What possible sources of error can you identify with this procedure? If you can think of ways to eliminate them, ask your teacher to approve your plan, and run the procedure again.

Students' suggestions for improving the procedure will vary. Possible

suggestions include performing repeated trials or using larger volumes.

Be sure answers are safe and include carefully planned procedures.

Name _____   Class _____   Date _____

Voltaic Cells

In voltaic cells, oxidation and reduction half-reactions take place in separate half-cells, which can consist of a metal electrode immersed in a solution of its metal ions. The electrical potential, or voltage, that develops between the electrodes is a measure of the combined reducing strength of one reactant and oxidizing strength of the other reactant.

OBJECTIVES

• **Construct** a Cu-Zn voltaic cell.

• **Design** and construct two other voltaic cells.

• **Measure** the potential of the voltaic cells.

• **Evaluate** cells by comparing the measured cell voltages with the voltages calculated from standard reduction potentials.

MATERIALS

• 0.5 M $Al_2(SO_4)_3$, 75 mL
• 0.5 M $CuSO_4$, 75 mL
• 0.5 M $ZnSO_4$, 75 mL
• Aluminum strip, 1 cm × 8 cm
• Copper strip, 1 cm × 8 cm
• Zinc strip, 1 cm × 8 cm
• Distilled water

• 100 mL graduated cylinder
• Emery cloth
• 150 mL beakers, 3
• Salt bridge
• Voltmeter
• Wires with alligator clips, 2

Always wear safety goggles and a lab apron to protect your eyes and clothing. If you get a chemical in your eyes, immediately flush the chemical out at the eyewash station while calling to your teacher. Know the locations of the emergency lab shower and the eyewash station and the procedures for using them.

Do not touch any chemicals. If you get a chemical on your skin or clothing, wash the chemical off at the sink while calling to your teacher. Make sure you carefully read the labels and follow the precautions on all containers of chemicals that you use. If there are no precautions stated on the label, ask your teacher what precautions you should follow. Do not taste any chemicals or items used in the laboratory. Never return leftovers to their original container; take only small amounts to avoid wasting supplies.

PREPARATION

1. Use the **Data Table** provided to record your data.

2. Remove any oxide coating from strips of aluminum, copper, and zinc by rubbing them with an emery cloth. Keep the metal strips dry until you are ready to use them.

3. Label three 150 mL beakers $Al_2(SO_4)_3$, $CuSO_4$, and $ZnSO_4$.

PROCEDURE

1. Pour 75 mL of 0.5 M $ZnSO_4$ into the $ZnSO_4$ beaker and 75 mL of 0.5 M $CuSO_4$ into the $CuSO_4$ beaker.

2. Place one end of the salt bridge into the $CuSO_4$ solution and the other end into the $ZnSO_4$ solution.

3. Place a zinc strip into the zinc solution and a copper strip into the copper solution.

4. Using the alligator clips, connect one wire to one end of the zinc strip and the second wire to the copper strip. Take the free end of the wire attached to the zinc strip, and connect it to one terminal on the voltmeter. Take the free end of the wire attached to the copper strip, and connect it to the other terminal on the voltmeter. The needle on the voltmeter should move to the right. If your voltmeter's needle points to the left, reverse the way the wires are connected to the terminals of the voltmeter. Immediately record the voltage reading in the **Data Table,** and disconnect the circuit.

5. Record the concentration of the solutions and sketch a diagram of your electrochemical cell.

6. Rinse the copper and zinc strips with a *very small* amount of distilled water. Collect the rinse from the copper strip in the $CuSO_4$ beaker and the rinse from the zinc strip in the $ZnSO_4$ beaker. Rinse each end of the salt bridge into the corresponding beaker.

7. Use the table of standard reduction potentials in the textbook to calculate the standard voltages for the other cells you can build using copper, zinc, or aluminum. Build these cells and measure their potentials following steps **1–6.**

DISPOSAL

8. Clean all apparatus and your lab station. Wash your hands. Place the pieces of metal in the containers designated by your teacher. Each solution should be poured in its own separate disposal container. Do not mix the contents of the beakers.

Name _____ Class _____ Date _____

Voltaic Cells *continued*

Data Table

Cell	$Zn|Zn^{2+}||Cu^{2+}|Cu$	$Al|Al^{3+}||Zn^{2+}|Zn$	$Al|Al^{3+}||Cu^{2+}|Cu$
Diagram			
Conc.	0.5 M Zn^{2+}	0.5 M Al^{3+}	0.5 M Al^{3+}
	0.5 M Cu^{2+}	0.5 M Zn^{2+}	0.5 M Cu^{2+}
Voltage (V)	0.96	0.83	1.90

Analysis

1. **Organizing Ideas** For each cell that you constructed, write the equations for the two half-cell reactions. Obtain the standard half-cell potentials for the half-reactions from the table in the textbook, and write these E^O values after the equations.

$Zn(s) \longrightarrow Zn^{2+}(aq) + 2e^-$
  $E^O = +0.76$ V

$Cu^{2+}(aq) + 2e^- \longrightarrow Cu(s)$
  $E^O = +0.34$ V

$Zn^{2+}(aq) + 2e^- \longrightarrow Zn(s)$
  $E^O = -0.76$ V

$Al(s) \longrightarrow Al^{3+}(aq) + 3e^-$
  $E^O = +1.66$ V

$Al(s) \longrightarrow Al^{3+}(aq) + 3e^-$
  $E^O = +1.66$ V

$Cu^{2+}(aq) + 2e^- \longrightarrow Cu(s)$
  $E^O = +0.34$ V

2. **Organizing Ideas** For each cell you tested, combine the two half-reactions to obtain the equation for the net reaction.

$Zn(s) + Cu^{2+}(aq) \longrightarrow Zn^{2+}(aq) + Cu(s)$

$2Al(s) + 3Zn^{2+}(aq) \longrightarrow 2Al^{3+}(aq) + 3Zn(s)$

$2Al(s) + 3Cu^{2+}(aq) \longrightarrow 2Al^{3+}(aq) + 3Cu(s)$

Voltaic Cells *continued*

3. **Organizing Ideas** Use the E^O values for the half-reactions to determine the E^O for each cell.

E^O **(Zn/Cu) = +0.76 V + 0.34 V = +1.10 V**

E^O **(Al/Cu) = −0.76 V + 1.66 V = +0.90 V**

E^O **(Al/Zn) = +1.66 V + 0.34 V = +2.00 V**

4. **Resolving Discrepancies** Compare the actual cell voltages you measured with the standard cell voltages in item 3. Explain why you would expect a difference.

Students will measure lower voltages than the standard electrode potentials

because they were not using 1.0 M solutions. The salt bridge is not a perfect

ion conductor, and polarization of the electrodes may occur.

Conclusions

1. **Inferring Conclusions** Based on the voltages that you measured, which cell produces the most energy?

The Al/Cu cell produces the most energy (2.00 V).

2. **Applying Ideas** On the basis of your data, which metal is the strongest reducing agent? Which metal ion is the strongest oxidizing agent?

Aluminum is the strongest reducing agent of the three metals. It is oxidized

by Zn^{2+} and Cu^{2+}. The copper(II) ion is the strongest oxidizing agent. It is

reduced by Al and Zn.

3. **Applying Ideas** Indicate the direction of electron flow in each of your cell diagrams.

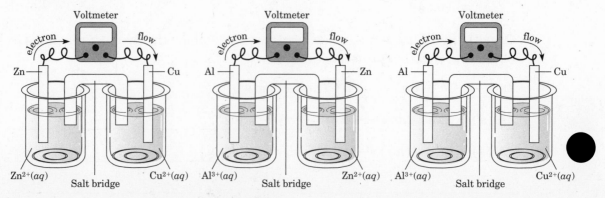

Voltaic Cells *continued*

EXTENSIONS

1. **Predicting Outcomes** Describe how and why the reactions would stop if the cells had been left connected.

   If the cells were left connected, they would continue producing an electric

   current until the amount of reactants decreases to a level that no longer

   drives the reaction to maintain an electric current.

   _____

2. **Designing Experiments** Design a method that could use several of the electrochemical cells you constructed to generate more voltage than any individual cell provided. (Hint: consider what would happen if you linked an Al-Zn cell and a Zn-Cu cell. If your teacher approves your plan, test your idea.)

   Students' answers will vary. The Al/Cu cell has the same potential as the

   Al/Zn cell plus the Zn/Cu cell. Students should realize that if they connect

   cells in series, they can increase the voltage. Be sure that answers are safe

   and include carefully planned procedures.

   _____

DATASHEET FOR IN-TEXT LAB

Simulation of Nuclear Decay Using Pennies and Paper

Radioactive isotopes are unstable. All radioactive matter decays, or breaks down, in a predictable pattern. Radioactive isotopes release radiation as they disintegrate into daughter isotopes.

The rate of decay is a measure of how fast an isotope changes into its daughter isotope. The rate of radioactive decay is conveniently characterized by the isotope's half-life, the period of time it takes one-half of the original material to decay. Half-lives vary from billions of years to fractions of a second.

OBJECTIVES

- **Infer** that the rate of decay can be simulated by a random process.
- **Compare** the numbers of pennies that are showing heads with the number showing tails.
- **Create** a string plot that represents nuclear decay.
- **Relate** observations to the rate of nuclear decay.
- **Graph** the data.
- **Compare** the results of the two simulation procedures.

MATERIALS

- colored paper or cloth strips, approximately 65 cm × 2.5 cm (2 strips)
- graph paper
- one sheet of stiff poster board, 70 cm × 60 cm
- pennies or other objects supplied by your teacher (100)
- scissors, tape, meter stick, pencil, and string
- shoe box with lid

Always wear safety goggles and a lab apron to protect your eyes and clothing. If you get a chemical in your eyes, immediately flush the chemical out at the eyewash station while calling to your teacher. Know the locations of the emergency lab shower and the eyewash station and the procedures for using them.

PREPARATION

For Part A, record your data in the **Data Table.**

PROCEDURE

Part A: Simulating radioactive decay with pennies

1. Place 100 pennies into the shoe box so that the head sides are up. The pennies will represent atoms. Record 100 in the "Unchanged atoms" column and 0 in the "Changed atoms" column.

| Simulation of Nuclear Decay Using Pennies and Paper *continued*

2. With the lid on the box, shake the box up and down 5 times. We will count each shaking period as being equivalent to 10 s.

3. Open the lid, and remove all of the pennies that have the tails side up. These pennies represent the changed atoms.

4. Count the number of pennies remaining in the box. Record this number in the 10 s row of the "Unchanged atoms" column in the **Data Table.** Count the number of changed atoms (the pennies that you removed from the box), and record the number in the 10 s row.

5. Each lab partner should predict how many times steps **2–4** will need to be repeated until only one unchanged atom remains. Record the time that each lab partner predicted. Remember that each shaking period is counted as 10 s, so four shaking periods would be recorded as 40 s.

6. Repeat steps **2–4** by counting and recording each time until only 1 (or 0) penny with the head side up remains.

Part B: Simulating decay with paper

7. Draw an y-axis and x-axis on the poster board so that they are about 5 cm from the left side and the bottom edge respectively. Label the x-axis as "Time" and the y-axis as "Amount of material."

8. Along the x-axis, draw marks every 10 cm from the y-axis line. Label the first mark "0" and the next mark "1," and so on. Each mark represents 1 minute.

9. Place one of the colored strips vertically with its lower edge centered on the 0 mark of the x-axis. Tape the strip in place.

10. Fold the other colored strip in half, and cut it in the middle. Place one-half of the strip so that it is centered on the next mark, and tape the strip in place.

11. Fold the remaining piece of the strip in half, and cut it exactly in the middle.

12. Place one of the pieces so that it is centered on the next mark, and tape the piece in place.

13. Repeat steps **11** and **12,** and each time, tape the first piece vertically at the next x-axis mark. Continue until you have at least 8 strips taped along the x-axis.

14. Use the string to join the tops of each strip of paper to make a continuous curve.

DISPOSAL

15. Return the pennies and box to your teacher. Dispose of the poster board, strips, and string as instructed by your teacher. Clean up your lab station.

| Simulation of Nuclear Decay Using Pennies and Paper *continued*

Data Table		
Time (s)	**Unchanged atoms**	**Changed atoms**
0	100	0
10	49	51
20	23	26
30	11	12
40	7	4
50	6	1
60	4	2
70	1	3

Analysis
PART A

1. **Predicting Outcomes** How long did it take to have only 1 penny (0 pennies) left in the box? How close was your prediction in step 5?

   Answers will vary. _____

   _____

2. **Analyzing Data** Make a graph of your data on a piece of graph paper. Label the x-axis "Time" and the y-axis "Unchanged atoms." Plot the number of unchanged atoms versus time. Draw a smooth curve through the data points.

   Plots of data will vary slightly, but the graphs should all display the same trend.

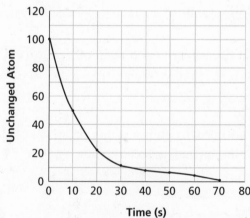

Penny Decay

Simulation of Nuclear Decay Using Pennies and Paper *continued*

3. **Analyzing Results** Each trial was comparable to a 10 s period of time. How long did it take for half of your pennies to be removed from the box? What is the half-life of the process?

   10 s; 10 s

4. **Interpreting Graphics** Use your graph to determine the time it takes to have only 25% of the unchanged atoms remaining. In your experiment, how many pennies remained in the box at that time?

   Answers will vary but should be 20 s or 30 s. Answers will vary but should be

   close to 25 pennies.

PART B

5. **Analyzing Results** How many half lives have passed after 4 minutes?

   4 half-lives

6. **Interpreting Graphics** Using the string plot, determine how many minutes it took until only 20% of the original material remained.

   Answers will vary but should be approximately 2.3 min.

Conclusions

1. **Inferring Conclusions** If you started with a paper strip that was twice as long, would the half-life change?

   The half-life will be the same.

2. **Inferring Conclusions** Is there a relationship between the graph from Part A and the string plot from Part B?

   Both curves have the same shape. In both cases, half of the original objects

   or half of the length disappears in each trial.

Skills Practice

Polymers and Toy Balls

What polymers make the best toy balls? Two possibilities are latex rubber and a polymer produced from ethanol and sodium silicate. Latex rubber is a polymer of covalently bonded atoms.

The polymer formed from ethanol, C_2H_5OH, and a solution of sodium silicate, $Na_2Si_3O_7$, also has covalent bonds. It is known as *water glass* because it dissolves in water.

In this experiment, you will synthesize rubber and the ethanol silicate polymer and test their properties.

OBJECTIVES

- **Synthesize** two different polymers.

- **Prepare** a small toy ball from each polymer.

- **Observe** the similarities and differences between the two types of balls.

- **Measure** the density of each polymer.

- **Compare** the bounce height of the two balls.

MATERIALS

- 2 L beaker, or plastic bucket or tub
- 3 mL 50% ethanol solution
- 5 oz paper cups, 2
- 10 mL 5% acetic acid solution (vinegar)
- 25 mL graduated cylinder
- 10 mL graduated cylinder

- 10 mL liquid latex
- 12 mL sodium silicate solution
- distilled water
- gloves
- meterstick
- paper towels
- wooden stick

Always wear safety goggles and a lab apron to protect your eyes and clothing. If you get a chemical in your eyes, immediately flush the chemical out at the eyewash station while calling to your teacher. Know the locations of the emergency lab shower and the eyewash station and the procedures for using them.

Do not touch any chemicals. If you get a chemical on your skin or clothing, wash the chemical off at the sink while calling to your teacher. Make sure you carefully read the labels and follow the precautions on all containers of chemicals that you use. If there are no precautions stated on the label, ask your teacher what precautions you should follow. Do not taste any chemicals or items used in the laboratory. Never return leftovers to their original container; take only small amounts to avoid wasting supplies.

Polymers and Toy Balls *continued*

PREPARATION

1. Organizing Data Use **Data Table 1** and **Data Table 2** to record your observations.

PROCEDURE

1. Fill the 2 L beaker, bucket, or tub about half-full with distilled water.

2. Using a clean 25 mL graduated cylinder, measure 10 mL of liquid latex and pour it into one of the paper cups.

3. Thoroughly clean the 25 mL graduated cylinder with soap and water, and then rinse it with distilled water.

4. Measure 10 mL of distilled water. Pour it into the paper cup with the latex.

5. Measure 10 mL of the 5% acetic acid solution, and pour it into the paper cup with the latex and water.

6. Immediately stir the mixture with the wooden stick.

7. As you continue stirring, a polymer lump will form around the wooden stick. Pull the stick with the polymer lump from the paper cup, and immerse the lump in the 2 L beaker, bucket, or tub.

8. While wearing gloves, gently pull the lump from the wooden stick. Be sure to keep the lump immersed under the water.

9. Keep the latex rubber underwater, and use your gloved hands to mold the lump into a ball. Then, squeeze the lump several times to remove any unused chemicals. You may remove the latex rubber from the water as you roll it in your hands to smooth the ball.

10. Set aside the latex rubber ball to dry. While it is drying, proceed to step **11.**

11. In a clean 25 mL graduated cylinder, measure 12 mL of sodium silicate solution, and pour it into the other paper cup.

12. In a clean 10 mL graduated cylinder, measure 3 mL of 50% ethanol. Pour the ethanol into the paper cup with the sodium silicate, and mix with the wooden stick until a solid substance is formed.

13. While wearing gloves, remove the polymer that forms and place it in the palm of one hand. Gently press it with the palms of both hands until a ball that does not crumble is formed. This step takes a little time and patience. The liquid that comes out of the ball is a combination of ethanol and water. Occasionally, moisten the ball by letting a small amount of water from a faucet run over it. When the ball no longer crumbles, you are ready to go to the next step.

14. Observe as many physical properties of the balls as possible, and record your observations in **Data Table 1** or **Data Table 2.**

15. Drop each ball several times, and record your observations.

Polymers and Toy Balls *continued*

16. Drop one ball from a height of 1 m, and measure its bounce. Perform three trials for each ball.

17. Measure the diameter and mass of each ball. Record these data in **Data Table 1** or **Data Table 2**.

DISPOSAL

18. Dispose of any extra solutions in the containers indicated by your teacher. Clean up your lab area. Remember to wash your hands thoroughly when your lab work is finished.

Data Table 1	
Latex rubber	
Bounce height—trial 1	**50 cm**
Bounce height—trial 2	**50 cm**
Bounce height—trial 3	**55 cm**
Mass	**45.20 g**
Diameter	**6.0 cm**
Other observations	**Does not crumble very easily. Seems more opaque.**

Data Table 2	
Ethanol-silicate polymer	
Bounce height—trial 1	**58 cm**
Bounce height—trial 2	**60 cm**
Bounce height—trial 3	**50 cm**
Mass	**84.03 g**
Diameter	**7.0 cm**
Other observations	**Tends to bounce higher than other balls. Crumbles easily. Seems translucent.**

Polymers and Toy Balls *continued*

ANALYSIS

1. Analyzing Information List at least three of your observations of the properties of the two balls.

Answers will vary but could include the following: The latex ball is more

opaque, less smooth, and less crumbly than the ethanol-silicate polymer ball.

The ethanol-silicate polymer ball breaks down after a period of time. Both

balls bounce. The ethanol-silicate ball bounces higher than the latex rubber

ball.

_____

2. Organizing Data Calculate the average height of the bounce for each type of ball.

Average bounce height for latex ball:

$$\frac{50 + 50 + 55}{3} = 52 \text{ cm}$$

Average bounce height for ethanol-silicate polymer ball:

$$\frac{58 + 60 + 50}{3} = 56 \text{ cm}$$

3. Organizing Data Calculate the volume for each ball. Even though the balls may not be perfectly spherical, assume that they are. (Hint: The volume of a sphere is equal to $\frac{4}{3} \times \pi \times r^3$, where r is the radius of the sphere, which is one-half of the diameter.) Then, calculate the density of each ball, using your mass measurements.

Volume of latex ball:

$$\frac{4}{3} = (3.14)(3.0 \text{ cm})^3 = 110 \text{ cm}^3$$

Volume of ethanol-silicate polymer ball:

$$\frac{4}{3} = (3.14)(3.5 \text{ cm})^3 = 180 \text{ cm}^3$$

Density of latex ball: $\dfrac{45.20 \text{ g}}{110 \text{ cm}^3} = 0.41 \text{ g/cm}^3$

Density of ethanol-silicate polymer ball: $\dfrac{84.03 \text{ g}}{180 \text{ cm}^3} = 0.47 \text{ g/cm}^3$

CONCLUSIONS

1. **Inferring Conclusions** Which polymer would you recommend to a toy company for making new toy balls? Explain your reasoning.

   Student answers will vary but should be based on the properties of the ball.

   Some students may argue that because the ethanol-silicate polymer ball

   crumbles in time, it would be a poor candidate for a toy.

2. **Evaluating Viewpoints** What are some other possible practical applications for each of the polymers you made?

   Student answers will vary but should be based on the properties of each

   substance.

EXTENSION

1. **Predicting Outcomes** Explain why you would not be able to measure the volumes of the balls by submerging them in water.

   The ethanol-silicate polymer ball is less dense than water, so it would float

   rather than being fully submerged. Also, it would break down in the water, so

   volume must be calculated from a measurement of diameter.

Name _____ Class _____ Date _____

Skills Practice

Casein Glue

Cow's milk contains averages of 4.4% fat, 3.8% protein, and 4.9% lactose. At the normal pH of milk, 6.3 to 6.6, the protein remains dispersed because it has a net negative charge due to the dissociation of the carboxylic acid group, as shown in Figure A below. As the pH is lowered by the addition of an acid, the protein acquires a net charge of zero, as shown in Figure B. After the protein loses its negative charge, it can no longer remain in solution, and it coagulates into an insoluble mass. The precipitated protein is known as casein and has a molecular mass between 75 000 and 375 000 amu. The pH at which the net charge on a protein becomes zero is called the isoelectric pH. For casein, the isoelectric pH is 4.6.

$$H_2N—\boxed{protein}—COO^- \qquad {}^+H_3N—\boxed{protein}—COO^-$$

Figure A **Figure B**

In this experiment, you will coagulate the protein in milk by adding acetic acid. The casein can then be separated from the remaining solution by filtration. This process is known as separating the curds from the whey. The excess acid in the curds can be neutralized by the addition of sodium hydrogen carbonate, $NaHCO_3$. The product of this reaction is casein glue. Do not eat or drink any materials or products of this lab.

OBJECTIVES

- **Recognize** the structure of a protein.
- **Predict** and **observe** the result of acidifying milk.
- **Prepare** and **test** a casein glue.
- **Deduce** the charge distribution in proteins as determined by pH.

MATERIALS

- 100 mL graduated cylinder
- 250 mL beaker
- 250 mL Erlenmeyer flask
- funnel
- glass stirring rod
- hot plate
- medicine dropper
- baking soda, $NaHCO_3$
- nonfat milk
- paper
- paper towel
- thermometer
- white vinegar
- wooden splints, 2

Casein Glue *continued*

Always wear safety goggles and a lab apron to protect your eyes and clothing. If you get a chemical in your eyes, immediately flush the chemical out at the eyewash station while calling to your teacher. Know the locations of the emergency lab shower and the eyewash station and the procedures for using them.

Do not touch any chemicals. If you get a chemical on your skin or clothing, wash the chemical off at the sink while calling to your teacher. Make sure you carefully read the labels and follow the precautions on all containers of chemicals that you use. If there are no precautions stated on the label, ask your teacher what precautions you should follow. Do not taste any chemicals or items used in the laboratory. Never return leftovers to their original container; take only small amounts to avoid wasting supplies.

PREPARATION

1. Use the **Data Table** provided for recording observations at each step of the procedure.

2. Predict the characteristics of the product that will be formed when the acetic acid is added to the milk. Record your predictions in the **Data Table.**

PROCEDURE

1. Pour 125 mL of nonfat milk into a 250 mL beaker. Add 20 mL of 4% acetic acid (white vinegar).

2. Place the mixture on a hot plate and heat it to 60°C. Record your observations in the **Data Table** and compare them with the predictions you made in Preparation step **2.**

3. Filter the mixture through a folded piece of paper towel into an Erlenmeyer flask.

4. Discard the filtrate, which contains the whey. Scrape the curds from the paper towel back into the 250 mL beaker.

5. Add 1.2 g of $NaHCO_3$ to the beaker and stir. Slowly add drops of water, stirring intermittently, until the consistency of white glue is obtained.

6. Use your glue to fasten together two pieces of paper. Also fasten together two wooden splints. Allow the splints to dry overnight, and then test the joint for strength.

DISPOSAL

7. Clean all apparatus and your lab station. Return equipment to its proper place. Dispose of chemicals and solutions in the containers designated by your teacher. Do not pour any chemicals down the drain or in the trash unless your teacher directs you to do so. Wash your hands thoroughly before you leave the lab and after all work is finished.

Data Table	
Predicted Result: milk + acetic acid	milk will coagulate
Actual Result milk: + acetic acid	milk coagulates

Analysis

1. **Organizing Ideas** Write the net ionic equation for the reaction between the excess acetic acid and the sodium hydrogen carbonate. Include the physical states of the reactants and products.

$$CH_3COOH(aq) + HCO_3^-(aq) \longrightarrow H_2O(l) + CH_3COO^-(aq) + CO_2(g)$$

2. **Evaluating Methods** In this experiment, what happened to the lactose and fat portions of the milk?

They are contained in the filtrate (whey) that is discarded.

Conclusion

1. **Inferring Conclusions** **Figure A** shows that the net charge on a protein is negative at pH values higher than its isoelectric pH because the carboxyl group is ionized. **Figure B** shows that at the isoelectric pH, the net charge is zero. Predict the net charge on a protein at pH values lower than the isoelectric point, and draw a diagram to represent the protein.

The charge is positive below the isoelectric pH.

^+H_3N—| protein |—COOH

Extensions

1. **Relating Ideas** **Figure B** represents a protein as a dipolar ion, or zwitterion. The charges in a zwitterion suggest that the carboxyl group donates a hydrogen ion to the amine group. Is there any other way to represent the protein in **Figure B** so that it still has a net charge of zero?

H_2N—| protein |—COOH

Casein Glue *continued*

2. **Designing Experiments** Design a strength-testing device for the glue joint between the two wooden splints. If your teacher approves your design, create the device and use it to test the strength of the glue.

Students' suggestions will vary. If students are to test their devices, make

sure that the procedures they propose are safe.

Extraction and Filtration

Extraction, the separation of substances in a mixture by using a solvent, depends on solubility. For example, sand can be separated from salt by adding water to the mixture. The salt dissolves in the water, and the sand settles to the bottom of the container. The sand can be recovered by decanting the water. The salt can then be recovered by evaporating the water.

Filtration separates substances based on differences in their physical states or in the size of their particles. For example, a liquid can be separated from a solid by pouring the mixture through a paper-lined funnel or, if the solid is more dense than the liquid, the solid will settle to the bottom of the container, leaving the liquid on top. The liquid can then be decanted, leaving the solid.

SETTLING AND DECANTING

1. Fill an appropriate-sized beaker with the solid-liquid mixture provided by your teacher. Allow the beaker to sit until the bottom is covered with solid particles and the liquid is clear.

2. Grasp the beaker with one hand. With the other hand, pick up a stirring rod and hold it along the lip of the beaker. Tilt the beaker slightly so that liquid begins to pour out in a slow, steady stream.

GRAVITY FILTRATION

1. Prepare a piece of filter paper. Fold it in half and then in half again. Tear the corner of the filter paper, and open the filter paper into a cone. Place it in the funnel.

2. Put the funnel, stem first, into a filtration flask, or suspend it over a beaker using an iron ring.

3. Wet the filter paper with distilled water from a wash bottle. The paper should adhere to the sides of the funnel, and the torn corner should prevent air pockets from forming between the paper and the funnel.

4. Pour the mixture to be filtered down a stirring rod into the filter. The stirring rod directs the mixture into the funnel and reduces splashing.

5. Do not let the level of the mixture in the funnel rise above the edge of the filter paper.

6. Use a wash bottle to rinse all of the mixture from the beaker into the funnel.

VACUUM FILTRATION

1. Check the T attachment to the faucet. Turn on the water. Water should run without overflowing the sink or spitting while creating a vacuum. To test for a vacuum, cover the opening of the horizontal arm of the T with your thumb or index finger. If you feel your thumb being pulled inward, you have a vacuum. Note the number of turns of the knob that are needed to produce the flow of water that creates a vacuum.

2. Turn the water off. Attach the pressurized rubber tubing to the *horizontal* arm of the T. (You do not want water to run through the tubing.)

3. Attach the free end of the rubber tubing to the side arm of a filter flask. Check for a vacuum. Turn on the water so that it rushes out of the faucet (refer to step 1). Place the palm of your hand over the opening of the **Erlenmeyer** flask. You should feel the vacuum pull your hand inward. If you do not feel any pull or if the pull is weak, increase the flow of water. If increasing the flow of water fails to work, shut off the water and make sure your tubing connections are tight.

4. Insert the neck of a Büchner funnel into a one-hole rubber stopper until the stopper is about two-thirds to three-fourths up the neck of the funnel. Place the funnel stem into the **Erlenmeyer** flask so that the stopper rests in the mouth of the flask.

5. Obtain a piece of round filter paper. Place it inside the Büchner funnel over the holes. Turn on the water as in step 1. Hold the filter flask with one hand, place the palm of your hand over the mouth of the funnel, and check for a vacuum.

6. Pour the mixture to be filtered into the funnel. Use a wash bottle to rinse all of the mixture from the beaker into the funnel.

Skills Practice)

Gravimetric Analysis

Gravimetric analytical methods are based on accurate and precise mass measurements. They are used to determine the amount or percentage of a compound or element in a sample material. For example, if we want to determine the percentage of iron in an ore or the percentage of chloride ion in drinking water, gravimetric analysis would be used.

A gravimetric procedure generally involves reacting the sample to produce a reaction product that can be used to calculate the mass of the element or compound in the original sample. For example, to calculate the percentage of iron in a sample of iron ore, the mass of the ore is determined. The ore is then dissolved in hydrochloric acid to produce $FeCl_3$. The $FeCl_3$ precipitate is converted to a hydrated form of Fe_2O_3 by adding water and ammonia to the system. The mixture is then filtered to separate the hydrated Fe_2O_3 from the mixture. The hydrated Fe_2O_3 is heated in a crucible to drive the water from the hydrate, producing anhydrous Fe_2O_3. The mass of the crucible and its contents is determined after successive heating steps to ensure that the product has reached constant mass and that all of the water has been driven off. The mass of Fe_2O_3 produced can be used to calculate the mass and percentage of iron in the original ore sample.

Gravimetric procedures require accurate and precise techniques and measurements to obtain suitable results. Possible sources of error are the following:

1. The product (precipitate) that is formed is contaminated.

2. Some product is lost when transferring the product from a filter to a crucible.

3. The empty crucible is not clean or is not at constant mass for the initial mass measurement.

4. The system is not heated sufficiently to obtain an anhydrous product.

Always wear safety goggles and a lab apron to protect your eyes and clothing. If you get a chemical in your eyes, immediately flush the chemical out at the eyewash station while calling to your teacher. Know the locations of the emergency lab shower and the eyewash station and the procedures for using them.

When using a Bunsen burner, confine long hair and loose clothing. Do not heat glassware that is broken, chipped, or cracked. Use tongs or a hot mitt to handle heated glassware and other equipment; heated glassware does not always look hot. If your clothing catches fire, WALK to the emergency lab shower and use it to put out the fire.

Never put broken glass or ceramics in a regular waste container. Broken glass and ceramics should be disposed of in a separate container designated by your teacher.

SETTING UP THE EQUIPMENT

1. Attach a metal ring clamp to a ring stand, and lay a clay triangle on the ring.

CLEANING THE CRUCIBLE

2. Wash and dry a metal or ceramic crucible and lid. Cover the crucible with its lid, and use a balance to obtain its mass. If the balance is located far from your working station, use crucible tongs to place the crucible and lid on a piece of wire gauze. Carry the crucible to the balance, using the wire gauze as a tray.

HEATING THE CRUCIBLE TO OBTAIN A CONSTANT MASS

3. After recording the mass of the crucible and lid, suspend the crucible over a Bunsen burner by placing it on the clay triangle. Then place the lid on the crucible so that the entire contents are covered.

4. Light the Bunsen burner. Heat the crucible for 5 minutes with a gentle flame, and then adjust the burner to produce a strong flame. Heat for 5 minutes more. Shut off the gas to the burner. Allow the crucible and lid to cool. Using crucible tongs, carry the crucible and lid to the balance. Measure and record the mass. If the mass differs from the mass before heating, repeat the process until mass data from heating trials are within 1% of each other. This assumes that the crucible has a constant mass. The crucible is now ready to be used in a gravimetric analysis procedure. Details will be found in the following experiments. Gravimetric methods are used in Experiment 7 to synthesize magnesium oxide, and to separate $SrCO_3$ from a solution in Experiment 9.

Name _____ Class _____ Date _____

DATASHEET FOR IN-TEXT LAB

Paper Chromatography

Chromatography is a technique used to separate substances dissolved in a mixture. The Latin roots of the word are *chromato*, which means "color," and *graphy*, which means "to write." Paper is one medium used to separate the components of a solution.

Paper is made of cellulose fibers that are pressed together. As a solution passes over the fibers and through the pores, the paper acts as a filter and separates the mixture's components. Particles of the same component group together, producing a colored band. Properties such as particle size, molecular mass, and charge of the different solute particles in the mixture affect the distance the components will travel on the paper. The components of the mixture that are the most soluble in the solvent and the least attracted to the paper will travel the farthest. Their band of color will be closest to the edge of the paper.

Always wear safety goggles and a lab apron to protect your eyes and clothing. If you get a chemical in your eyes, immediately flush the chemical out at the eyewash station while calling to your teacher. Know the location of the emergency lab shower and the eyewash station and the procedures for using them.

PROCEDURE

1. Use a lead pencil to sketch a circle about the size of a quarter in the center of a piece of circular filter paper that is 12 cm in diameter.

2. Write one numeral for each substance, including any unknowns, around the inside of this circle. In this experiment, 6 mixtures are to be separated, so the circle is labeled 1 through 6, as shown in **Figure A.**

3. Use a micropipet to place a spot of each substance to be separated next to a number. Make one spot per number. If the spot is too large, you will get a broad, tailing trace with little or no detectable separation.

4. Use the pencil to poke a small hole in the center of the spotted filter paper. Insert a wick through the hole. A wick can be made by rolling a triangular piece of filter paper into a cylinder: start at the point of the triangle, and roll toward its base.

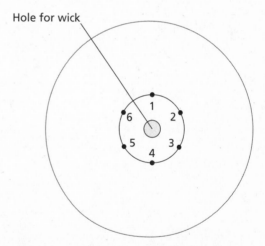

Hole for wick

Figure A Filter paper used in paper chromatography is spotted with the mixtures to be separated. Each spot is labeled with a numeral or a name that identifies the mixture to be separated. A hole punched in the center of the paper will attach to a wick that delivers the solvent to the paper.

T139 Teacher's Guide: Pre-Laboratory Procedure

5. Fill a petri dish or lid two-thirds full of solvent (usually water or alcohol).

6. Set the bottom of the wick in the solvent so that the filter paper rests on the top of the petri dish.

7. When the solvent is 1 cm from the outside edge of the paper, remove the paper from the petri dish, and allow the chromatogram to dry.

Most writing or drawing inks are mixtures of various components that give them specific properties. Therefore, paper chromatography can be used to study the composition of an ink. Experiment 12 investigates the composition of ball-point pen ink.

DATASHEET FOR IN-TEXT LAB

Volumetric Analysis

Volumetric analysis, the quantitative determination of the concentration of a solution, is achieved by adding a substance of known concentration to a substance of unknown concentration until the reaction between them is complete. The most common application of volumetric analysis is titration.

A buret is used in titrations. The solution with the known concentration is usually in the buret. The solution with the unknown concentration is usually in the Erlenmeyer flask. A few drops of a visual indicator also are added to the flask. The solution in the buret is then added to the flask until the indicator changes color, signaling that the reaction between the two solutions is complete. Then, using the volumetric data obtained and the balanced chemical equation for the reaction, the unknown concentration is calculated.

Always wear safety goggles and a lab apron to protect your eyes and clothing. If you get a chemical in your eyes, immediately flush the chemical out at the eyewash station while calling to your teacher. Know the locations of the emergency lab shower and the eyewash station and the procedures for using them.

ASSEMBLING THE APPARATUS

1. Attach a buret clamp to a ring stand.

2. Thoroughly wash and rinse a buret. If water droplets cling to the walls of the buret, wash it again and gently scrub the inside walls with a buret brush.

3. Attach the buret to one side of the buret clamp.

4. Place a Erlenmeyer flask for waste solutions under the buret tip.

OPERATING THE STOPCOCK

1. The stopcock should be operated with the left hand. This method gives better control but may prove awkward at first for right-handed students. The handle should be moved with the thumb and first two fingers of the left hand.

2. Rotate the stopcock back and forth. It should move freely and easily. If it sticks or will not move, ask your teacher for assistance. Turn the stopcock to the closed position. Use a wash bottle to add 10 mL of distilled water to the buret. Rotate the stopcock to the open position. The water should come out in a steady stream. If no water comes out or if the stream of water is blocked, ask your teacher to check the stopcock for clogs.

FILLING THE BURET

1. To fill the buret, place a funnel in the top of the buret. Slowly and carefully pour the solution of known concentration from a beaker into the funnel. Open the stopcock, and allow some of the solution to drain into the waste beaker. Then add enough solution to the buret to raise the level above the zero mark, but do not allow the solution to overflow.

Volumetric Analysis *continued*

READING THE BURET

1. Drain the buret until the bottom of the meniscus is on the zero mark or within the calibrated portion of the buret. If the solution level is not at zero, record the exact reading. If you start from the zero mark, your final buret reading will equal the amount of solution added. Remember, burets can be read to the second decimal place. Burets are designed to read the volume of liquid delivered to the flask, so numbers increase as you read downward from the top. .

2. Replace the waste beaker with an Erlenmeyer flask containing a measured amount of the solution of unknown concentration.

Experiment 15-1 is an example of a back-titration applied to an acid-base reaction; it can be performed on a larger scale if micropipets are replaced with burets.

Skills Practice

Calorimetry

Calorimetry, the measurement of the transfer of energy as heat, allows chemists to determine thermal constants, such as the specific heat of metals and the enthalpy of solution.

When two substances at different temperatures touch one another, energy as heat flows from the warmer substance to the cooler substance until the two substances are at the same temperature. The amount of energy transferred is measured in joules. (One joule equals 4.184 calories.)

A device used to measure the transfer of energy as heat is a calorimeter. Calorimeters vary in construction depending on the purpose and the accuracy of the energy measurement required. No calorimeter is a perfect insulator; some energy is always lost to the surroundings as heat. Therefore, every calorimeter must be calibrated to obtain its calorimeter constant.

Always wear safety goggles and a lab apron to protect your eyes and clothing. If you get a chemical in your eyes, immediately flush the chemical out at the eyewash station while calling to your teacher. Know the locations of the emergency lab shower and the eyewash station and the procedures for using them.

Turn off hot plates and other heat sources when not in use. Do not touch a hot plate after it has just been turned off; it is probably hotter than you think. Use tongs when handling heated containers. Never hold or touch containers with your hands while heating them.

The steps for constructing a calorimeter made from plastic foam cups follow.

CONSTRUCTING THE CALORIMETER

1. Trim the lip of one plastic foam cup, and use that cup as the top of your calorimeter. The other cup will be used as the base.

2. Use the pointed end of a pencil to gently make a hole in the center of the calorimeter top. The hole should be large enough to insert a thermometer. Make a hole for the wire stirrer. This hole should be positioned so that the wire stirrer can be raised and lowered without interfering with the thermometer.

3. Place the calorimeter in a beaker to prevent it from tipping over.

CALIBRATING A PLASTIC FOAM CUP CALORIMETER

1. Measure 50 mL of distilled water in a graduated cylinder. Pour it into the calorimeter. Measure and record the temperature of the water in the polystyrene cup.

2. Pour another 50 mL of distilled water into a beaker. Set the beaker on a hot plate, and warm the water to about 60°C. Measure and record the temperature of the water.

Name _____ Class _____ Date _____

Calorimetry *continued*

3. Immediately pour the warm water into the cup. Cover the cup, and move the stirrer gently up and down to mix the contents thoroughly. **Take care not to break the thermometer.**

4. Watch the thermometer and record the highest temperature attained (usually after about 30 s).

5. Empty the calorimeter.

6. The derivation of the equation to find the calorimeter constant starts with the following relationship.

Energy lost by the warm water = Energy gained by the cool water + Energy gained by the calorimeter

$$q_{warm\ H_2O} = q_{cool\ H_2O} + q_{calorimeter}$$

The energy lost as heat by the warm water is calculated by

$$q_{warm\ H_2O} = mass_{warm\ H_2O} \times 4.184\ J/(g \cdot °C) \times \Delta t$$

The energy gained as heat by the calorimeter system equals the energy lost as heat by the warm water. You can use the following equation to calculate the calorimeter constant C′ for your calorimeter.

$$q_{calorimeter} = q_{warm\ H_2O} = (mass_{cool\ H_2O})\,(4.184\ J/g \times °C)\,(\Delta t_{cool\ H_2O}) +$$
$$C'\,(\Delta t_{cool\ H_2O})$$

Substitute the data from your calibration and solve for C′.

(Inquiry)
Mixture Separation

The ability to separate and recover pure substances from mixtures is extremely important in scientific research and industry. Chemists need to work with pure substances, but naturally occurring materials are seldom pure. Often, differences in the physical properties of the components in a mixture provide the means for separating them. In this experiment, you will have an opportunity to design, develop, and implement your own procedure for separating a mixture. The mixture you will work with contains salt, sand, iron filings, and poppy seeds. All four substances are in dry, granular form.

OBJECTIVES

• **Observe** the chemical and physical properties of a mixture.

• **Relate** knowledge of chemical and physical properties to the task of purifying the mixture.

• **Analyze** the success of methods of purifying the mixture.

MATERIALS

• aluminum foil

• cotton balls

• distilled water

• filter funnels

• filter paper

• forceps

• magnet

• paper clips

• paper towels

• Petri dish

• pipets

• plastic forks

• plastic spoons

• plastic straws

• rubber stoppers

• sample of mixture and components
(sand, iron filings, salt, poppy seeds)

• test tubes and rack

• tissue paper

• transparent tape

• wooden splints

Always wear safety goggles and a lab apron to protect your eyes and clothing. If you get a chemical in your eyes, immediately flush the chemical out at the eyewash station while calling to your teacher. Know the locations of the emergency lab shower and the eyewash station and the procedures for using them.

PREPARATION

Your task will be to plan and carry out the separation of a mixture. Before you can plan your experiment, you will need to investigate the properties of each component in the mixture. The properties will be used to design your mixture separation.

Mixture Separation *continued*

PROCEDURE

1. Obtain separate samples of each of the four mixture components from your teacher. Use the equipment you have available to make observations of the components and determine their properties. You will need to run several tests with each substance, so don't use all of your sample on the first test. Look for things like whether the substance is magnetic, whether it dissolves, or whether it floats. Record your observations in the **Data Table.**

2. Make a plan for what you will do to separate a mixture that includes the four components from step **1.** Review your plan with your teacher.

3. Obtain a sample of the mixture from your teacher. Using the equipment you have available, run the procedure you have developed.

DISPOSAL

4. Clean your lab station. Clean all equipment, and return it to its proper place. Dispose of chemicals and solutions in the containers designated by your teacher. Do not pour any chemicals down the drain or throw anything in the trash unless your teacher directs you to do so. Wash your hands thoroughly after all work is finished and before you leave the lab.

Data Table				
Properties	**Sand**	**Iron filings**	**Salt**	**Poppy seeds**
Dissolves				
Floats				
Magnetic				
Other				

Analysis

1. **Evaluating Methods** On a scale of 1 to 10, how successful were you in separating and recovering each of the four components: sand, salt, iron filings, and poppy seeds? Consider 1 to be the best and 10 to be the worst. Justify your ratings based on your observations.

_____

_____

_____

_____

_____

Conclusions

1. Evaluating Methods How did you decide on the order of your procedural steps? Would any order have worked?

_____

_____

_____

2. Designing Experiments If you could do the lab over again, what would you do differently? Be specific.

_____

_____

_____

_____

3. Designing Experiments Name two materials or tools that weren't available that might have made your separation easier.

_____

_____

_____

_____

4. Applying Ideas: For each of the four components, describe a specific physical property that enabled you to separate the component from the rest of the mixture.

_____

_____

_____

_____

_____

Mixture Separation *continued*

EXTENSIONS

1. Evaluating Methods What methods could be used to determine the purity of each of your recovered components?

_____

_____

_____

_____

_____

_____

2. Designing Experiments How could you separate each of the following two-part mixtures?

a. aluminum filings and iron filings

_____

_____

b. sand and gravel

_____

_____

c. sand and finely ground polystyrene foam

_____

_____

d. salt and sugar

_____

_____

_____

e. alcohol and water

_____

_____

f. nitrogen and oxygen

_____

_____

Mixture Separation *continued*

3. Designing Experiments One of the components of the mixture in this experiment is in a different physical state at the completion of this experiment than it was at the start. Which one? How would you convert that component back to its original state?

_____

_____

_____

_____

_____

Quick Lab

Density of Pennies

MATERIALS

• balance

• 100 mL graduated cylinder

• 40 pennies dated before 1982

• 40 pennies dated after 1982

• water

Always wear safety goggles and a lab apron to protect your eyes and clothing. If you get a chemical in your eyes, immediately flush the chemical out at the eyewash station while calling to your teacher. Know the locations of the emergency lab shower and the eyewash station and the procedures for using them.

PROCEDURE

1. Using the balance, determine the mass of the 40 pennies minted prior to 1982. Repeat this measurement two more times. Average the results of the three trials to determine the average mass of the pennies.

2. Repeat step **1** with the 40 pennies minted after 1982.

3. Pour about 50 mL of water into the 100 mL graduated cylinder. Record the exact volume of the water. Add the 40 pennies minted before 1982. CAUTION: Add the pennies carefully so that no water is splashed out of the cylinder. Record the exact volume of the water and pennies. Repeat this process two more times. Determine the volume of the pennies for each trial. Average the results of those trials to determine the average volume of the pennies.

4. Repeat step **3** with the 40 pennies minted after 1982.

5. Review your data for any large differences between trials that could increase the error of your results. Repeat those measurements.

6. Use the average volume and average mass to calculate the average density for each group of pennies.

7. Compare the calculated average densities with the density of the copper listed in **Table 4** on page 38 of the textbook.

DISCUSSION

1. Why is it best to use the results of three trials rather than a single trial for determining the density?

_____

_____

_____

Density of Pennies *continued*

2. How did the densities of the two groups of pennies compare? How do you account for any difference?

_____

_____

_____

3. Use the results of this investigation to formulate a hypothesis about the composition of the two groups of pennies. How could you test your hypothesis?

_____

_____

_____

_____

_____

Percentage of Water in Popcorn

Popcorn pops because of the natural moisture inside each kernel. When the internal water is heated above 100°C, the liquid water changes to a gas, which takes up much more space than the liquid, so the kernel expands rapidly.

    The percentage of water in popcorn can be determined by the following equation.

$$\frac{initial\ mass - final\ mass}{initial\ mass} \times 100 = percentage\ of\ H_2O\ in\ unpopped\ popcorn$$

The popping process works best when the kernels are first coated with a small amount of vegetable oil. Make sure you account for the presence of this oil when measuring masses. In this lab, you will design a procedure for determining the percentage of water in three samples of popcorn. The popcorn is for testing only, and *must not* be eaten.

OBJECTIVES

• **Measure** the masses of various combinations of a beaker, oil, and popcorn kernels.

• **Determine** the percentages of water in popcorn kernels.

• **Compare** experimental data.

MATERIALS

• aluminum foil (1 sheet)

• beaker, 250 mL

• Bunsen burner with gas tubing and striker

• kernels of popcorn for each of three brands (80)

• oil to coat the bottom of the beaker

• ring stand, iron ring, and wire gauze

    Always wear safety goggles and a lab apron to protect your eyes and clothing. If you get a chemical in your eyes, immediately flush the chemical out at the eyewash station while calling to your teacher. Know the locations of the emergency lab shower and the eyewash station and the procedures for using them.

    When using a Bunsen burner, confine long hair and loose clothing. If your clothing catches on fire, WALK to the emergency lab shower and use it to put out the fire. When heating a substance in a test tube, the mouth of the test tube should point away from where you and others are standing. Watch the test tube at all times to prevent the contents from boiling over.

PREPARATION

Use the **Data Table** provided to record your data.

PROCEDURE

1. Measure the mass of a 250 mL beaker. Record the mass in the **Data Table.**

2. Add a small amount of vegetable oil to the beaker to coat the bottom of it. Measure the mass of the beaker and oil. Record the mass in the **Data Table.**

3. Add 20 kernels of brand A popcorn to the beaker. Shake the beaker gently to coat the kernels with oil. Measure the mass of the beaker, oil, and popcorn. Record the mass in the **Data Table.**

4. Subtract the mass found in step **2** from the mass found in step **3** to obtain the mass of 20 unpopped kernels. Record the mass in the **Data Table.**

5. Cover the beaker loosely with the aluminum foil. Punch a few small holes in the aluminum foil to let moisture escape. These holes should not be large enough to let the popping corn pass through.

6. Heat the popcorn until the majority of the kernels have popped. The popcorn pops more efficiently if the beaker is held firmly with tongs and gently shaken side to side on the wire gauze.

7. Remove the aluminum foil from the beaker and allow the beaker to cool for 10 minutes. Then, measure the mass of the beaker, oil, and popped corn. Record the mass in the **Data Table.**

8. Subtract the mass in step **7** from the mass in step **3** to obtain the mass of water that escaped when the corn popped. Record the mass in the **Data Table.**

9. Calculate the percentage of water in the popcorn.

10. Dispose of the popcorn in the designated container. Remove the aluminum foil, and set it aside. Clean the beaker, and dry it well. Alternatively, if your teacher approves, use a different 250 mL beaker.

11. Repeat steps **1–10** for brand B popcorn.

12. Repeat steps **1–10** for brand C popcorn.

DISPOSAL

13. Dispose of popped popcorn and aluminum foil in containers as directed by your instructor. Do not eat the popcorn.

14. Clean beakers. Return beakers and other equipment to the proper place.

15. Clean all work surfaces and personal protective equipment as directed by your instructor.

16. Wash your hands thoroughly before leaving the laboratory.

Percentage of Water in Popcorn *continued*

Data Table			
	Popcorn Brand A	**Popcorn Brand B**	**Popcorn Brand C**
Mass of 250 mL beaker (g)			
Mass of beaker + oil (g)			
Mass of beaker + oil + 20 kernels (before) (g)			
Mass of 20 kernels (before) (g)			
Mass of beaker + oil + 20 kernels (after) (g)			
Mass of 20 kernels (after) (g)			
Mass of water in 20 kernels (g)			
Percentage of water in popcorn			

Analysis

1. Applying Ideas: Determine the mass of the 20 unpopped kernels of popcorn for each brand of popcorn.

_____

_____

_____

2. Applying Ideas: Determine the mass of the 20 popped kernels of popcorn for each brand of popcorn.

_____

_____

_____

3. Applying Ideas: Determine the mass of the water that was lost when the popcorn popped for each brand.

_____

_____

_____

▌Percentage of Water in Popcorn *continued*

Conclusions

1. Applying Data: Determine the percentage by mass of water in each brand of popcorn.

_____

_____

_____

_____

_____

2. Inferring Relationships: Do all brands of popcorn contain the same percentage water?

_____

_____

_____

_____

EXTENSIONS

1. Designing Experiments: What are some likely areas of imprecision in this experiment?

_____

_____

_____

_____

_____

_____

_____

_____

2. **Designing Experiments:** Do you think that the volume of popped corn depends on the percentage of water in the unpopped corn? Design an experiment to find the answer.

_____

_____

_____

_____

_____

_____

_____

_____

_____

_____

Name _____ Class _____ Date _____

Constructing a Model

How can you construct a model of an unknown object by (1) making inferences about an object that is in a closed container and (2) touching the object without seeing it?

MATERIALS

- can covered by a sock sealed with tape
- one or more objects that fit in the container
- metric ruler
- balance

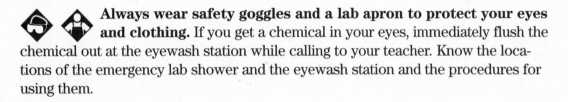 **Always wear safety goggles and a lab apron to protect your eyes and clothing.** If you get a chemical in your eyes, immediately flush the chemical out at the eyewash station while calling to your teacher. Know the locations of the emergency lab shower and the eyewash station and the procedures for using them.

PROCEDURE

Record all of your results in the **Data Table.**

1. Your teacher will provide you with a can that is covered by a sock sealed with tape. Without unsealing the container, try to determine the number of objects inside the can as well as the mass, shape, size, composition, and texture of each. To do this, you may carefully tilt or shake the can. Record your observations in the **Data Table.**

2. Remove the tape from the top of the sock. Do not look inside the can. Put one hand through the opening, and make the same observations as in step **1** by handling the objects. To make more-accurate estimations, practice estimating the sizes and masses of some known objects outside the can. Then compare your estimates of these objects with actual measurements using a metric ruler and a balance.

Data Table	
Observations	
Sealed can	**Unsealed can**

Constructing a Model *continued*

DISCUSSION

1. Scientists often use more than one method to gather data. How was this illustrated in the investigation?

_____

_____

_____

_____

_____

_____

2. Of the observations you made, which were qualitative and which were quantitative?

_____

_____

_____

3. Using the data you gathered, draw a model of the unknown object(s) and write a brief summary of your conclusions.

_____

_____

_____

_____

DATASHEET FOR IN-TEXT LAB

Conservation of Mass

The law of conservation of mass states that matter is neither created nor destroyed during a chemical reaction. Therefore, the mass of a system should remain constant during any chemical process. In this experiment, you will determine whether mass is conserved by examining a simple chemical reaction and comparing the mass of the system before the reaction with its mass after the reaction.

OBJECTIVES

• **Observe** the signs of a chemical reaction.

• **Compare** masses of reactants and products.

• **Design** experiments.

• **Relate** observations to the law of conservation of mass.

MATERIALS

• 2 L plastic soda bottle

• 5% acetic acid solution (vinegar)

• balance

• clear plastic cups, 2

• graduated cylinder

• hook-insert cap for bottle

• microplunger

• sodium hydrogen carbonate (baking soda)

  Always wear safety goggles and a lab apron to protect your eyes and clothing. If you get a chemical in your eyes, immediately flush the chemical out at the eyewash station while calling to your teacher. Know the locations of the emergency lab shower and the eyewash station and the procedures for using them.

PREPARATION

Use the data tables provided to record your data and observations for Part I and Part II.

PROCEDURE—PART I

1. Obtain a microplunger, and tap it down into a sample of baking soda until the bulb end is packed with a plug of the powder (4–5 mL of baking soda should be enough to pack the bulb).

2. Hold the microplunger over a plastic cup, and squeeze the sides of the microplunger to loosen the plug of baking soda so that it falls into the cup.

3. Use a graduated cylinder to measure 100 mL of vinegar, and pour it into a second plastic cup.

Conservation of Mass *continued*

4. Place the two cups side by side on the balance pan, and measure the total mass of the system (before reaction) to the nearest 0.01 g. Record the mass in **Data Table-Part I.**

5. Add the vinegar to the baking soda a little at a time to prevent the reaction from getting out of control. Allow the vinegar to slowly run down the inside of the cup. Observe and record your observations about the reaction.

6. When the reaction is complete, place both cups on the balance, and determine the total final mass of the system to the nearest 0.01 g. Calculate any change in mass. Record both the final mass and any change in mass in **Data Table-Part I.**

7. Examine the plastic bottle and the hook-insert cap. Try to develop a modified procedure that will test the law of conservation of mass more accurately than the procedure in Part I.

8. Write the answers to items 1 through 3 in Analysis—Part I.

PROCEDURE—PART II

9. Your teacher should approve the procedure you designed in Procedure—Part I, step **7.** Implement your procedure with the same chemicals and quantities you used in Part I, but use the bottle and hook-insert cap in place of the two cups. Record your data in **Data Table-Part II.**

10. If you were successful in step **9** and your results reflect the conservation of mass, proceed to complete the experiment. If not, find a lab group that was successful, and discuss with them what they did and why they did it. Your group should then test the other group's procedure to determine whether their results are reproducible.

DISPOSAL

11. Clean your lab station. Clean all equipment, and return it to its proper place. Dispose of chemicals and solutions in the containers designated by your teacher. Do not pour any chemicals down the drain or throw anything in the trash unless your teacher directs you to do so. Wash your hands thoroughly after all work is finished and before you leave the lab.

Conservation of Mass *continued*

Data Table-Part I			
Initial mass (g)	Final mass (g)	Change in mass (g)	Observations

Data Table-Part II			
Initial mass (g)	Final mass (g)	Change in mass (g)	Observations

Analysis

PART I

1. Drawing Conclusions What evidence was there that a chemical reaction occurred?

_____

_____

_____

2. Organizing Data How did the final mass of the system compare with the initial mass of the system?

_____

_____

_____

_____

Conservation of Mass *continued*

3. Resolving Discrepancies Does your answer to the previous question show that the law of conservation of mass was violated? (Hint: Another way to express the law of conservation of mass is to say that the mass of all of the products equals the mass of all of the reactants.) What do you think might cause the mass difference?

_____

_____

_____

_____

_____

Analysis
PART II

1. Drawing Conclusions Was there any new evidence in Part II indicating that a chemical reaction occurred?

_____

_____

_____

_____

_____

2. Organizing Ideas Identify the state of matter for each reactant in Part II. Identify the state of matter for each product.

_____

_____

_____

_____

_____

| Conservation of Mass *continued*

Conclusions

1. **Relating Ideas** What is the difference between the system in Part I and the system in Part II? What change led to the improved results in Part II?

_____

_____

_____

2. **Evaluating Methods** Why did the procedure for Part II work better than the procedure for Part I?

_____

_____

_____

_____

EXTENSIONS

1. **Applying Models** When a log burns, the resulting ash obviously has less mass than the unburned log did. Explain whether this loss of mass violates the law of conservation of mass.

_____

_____

_____

_____

_____

_____

_____

_____

Conservation of Mass *continued*

2. **Designing Experiments** Design a procedure that would test the law of conservation of mass for the burning log described in Extension item 1.

_____

_____

_____

_____

_____

_____

_____

_____

_____

_____

Quick Lab)

The Wave Nature of Light: Interference

Does light show the wave property of interference when a beam of light is projected through a pinhole onto a screen?

MATERIALS

- scissors
- manila folders
- thumbtack
- masking tape

- aluminum foil
- white poster board or cardboard
- flashlight

PROCEDURE

Record all your observations.

1. To make the pinhole screen, cut a 20 cm × 20 cm square from a manila folder. In the center of the square, cut a 2 cm square hole. Cut a 7 cm × 7 cm square of aluminum foil. Using a thumbtack, make a pinhole in the center of the foil square. Tape the aluminum foil over the 2 cm square hole, making sure the pinhole is centered.

2. Use white poster board to make a projection screen 35 cm × 35 cm.

3. In a dark room, center the light beam from a flashlight on the pinhole. Hold the flashlight about 1 cm from the pinhole. The pinhole screen should be about 50 cm from the projection screen. Adjust the distance to form a sharp image on the projection screen.

DISCUSSION

1. Did you observe interference patterns on the screen?

_____

_____

_____

_____

2. As a result of your observations, what do you conclude about the nature of light?

_____

_____

Flame Tests

The characteristic light emitted by an element is the basis for the chemical test known as a *flame test*.

To identify an unknown substance, you must first determine the characteristic colors produced by different elements. You will do this by performing a flame test on a variety of standard solutions of metal compounds. Then, you will perform a flame test with an unknown sample to see if it matches any of the standard solutions. The presence of even a speck of another substance can interfere with the identification of the true color of a particular type of atom, so be sure to keep your equipment very clean and perform multiple trials to check your work.

OBJECTIVES

- **Identify** a set of flame-test color standards for selected metal ions.

- **Relate** the colors of a flame test to the behavior of excited electrons in a metal ion.

- **Identify** an unknown metal ion by using a flame test.

- **Demonstrate** proficiency in performing a flame test and in using a spectroscope.

MATERIALS

- 250 mL beaker
- Bunsen burner and related equipment
- cobalt glass plates
- crucible tongs
- distilled water
- flame-test wire
- glass test plate (or a microchemistry plate with wells)
- spectroscope

- 1.0 M HCl solution
- $CaCl_2$ solution
- K_2SO_4 solution
- Li_2SO_4 solution
- Na_2SO_4 solution
- $SrCl_2$ solution
- unknown solution
- wooden splints (optional)

Always wear safety goggles and a lab apron to protect your eyes and clothing. If you get a chemical in your eyes, immediately flush the chemical out at the eyewash station while calling to your teacher. Know the locations of the emergency lab shower and the eyewash station and the procedures for using them.

Flame Tests *continued*

Do not touch any chemicals. If you get a chemical on your skin or clothing, wash the chemical off at the sink while calling to your teacher. Make sure you carefully read the labels and follow the precautions on all containers of chemicals that you use. If there are no precautions stated on the label, ask your teacher what precautions you should follow. Do not taste any chemicals or items used in the laboratory. Never return leftovers to their original container; take only small amounts to avoid wasting supplies.

When using a Bunsen burner, confine long hair and loose clothing. If your clothing catches on fire, WALK to the emergency lab shower and use it to put out the fire. When heating a substance in a test tube, the mouth of the test tube should point away from where you and others are standing. Watch the test tube at all times to prevent the contents from boiling over.

PREPARATION

1. Use the **Data Table** provided to record your data.

2. Label a beaker *Waste*. Thoroughly clean and dry a well strip. Fill the first well one-fourth full with 1.0 M HCl on the plate. Clean the test wire by first dipping it in the HCl and then holding it in the colorless flame of the Bunsen burner. Repeat this procedure until the flame is not colored by the wire. When the wire is ready, rinse the well with distilled water and collect the rinse water in the waste beaker.

3. Put 10 drops of each metal ion solution listed in the materials list in a row in each well of the well strip. Put a row of 1.0 M HCl drops on a glass plate across from the metal ion solutions. Record the positions of all of the chemicals placed in the wells. The wire will need to be cleaned thoroughly between each test solution with HCl to avoid contamination from the previous test.

PROCEDURE

1. Dip the wire into the $CaCl_2$ solution, and then hold it in the Bunsen burner flame. Observe the color of the flame, and record it in the **Data Table.** Repeat the procedure again, but this time look through the spectroscope to view the results. Record the wavelengths you see from the flame. Repeat each test three times. Clean the wire with the HCl as you did in Preparation step **2.**

2. Repeat step **1** with the K_2SO_4 and with each of the remaining solutions in the well strip.

Flame Tests *continued*

3. Test another drop of Na_2SO_4, but this time view the flame through two pieces of cobalt glass. Clean the wire, and repeat the test. Record in the **Data Table** the colors and wavelengths of the flames as they appear when viewed through the cobalt glass. Clean the wire and the well strip, and rinse the well strip with distilled water. Pour the rinse water into the waste beaker.

4. Put a drop of K_2SO_4 in a clean well. Add a drop of Na_2SO_4. Perform a flame test for the mixture. Observe the flame without the cobalt glass. Repeat the test again, but this time observe the flame through the cobalt glass. Record in the **Data Table** the colors and wavelengths of the flames. Clean the wire, and rinse the well strip with distilled water. Pour the rinse water into the waste beaker.

5. Obtain a sample of the unknown solution. Perform flame tests for it with and without the cobalt glass. Record your observations. Clean the wire, and rinse the well strip with distilled water. Pour the rinse water into the waste beaker.

DISPOSAL

6. Dispose of the contents of the waste beaker in the container designated by your teacher. Wash your hands thoroughly after cleaning up the area and equipment.

Data Table		
Metal compound	**Color of flame**	**Wavelengths detected (nm)**
$CaCl_2$		
K_2SO_4		
Li_2SO_4		
Na_2SO_4		
$SrCl_2$		
Na only (cobalt glass)		
K only (cobalt glass)		
Na and K		
Na and K (cobalt glass)		
Unknown		

Flame Tests *continued*

Analysis

1. **Organizing Data** Examine your data table, and create a summary of the flame test for each metal ion.

_____

_____

_____

_____

2. **Analyzing Data** Account for any differences in the individual trials for the flame tests for the metals ions.

_____

_____

_____

_____

_____

3. **Organizing Ideas** Explain how viewing the flame through cobalt glass can make it easier to analyze the ions being tested.

_____

_____

_____

_____

4. **Relating Ideas** For three of the metal ions tested, explain how the flame color you saw relates to the lines of color you saw when you looked through the spectroscope.

_____

_____

_____

_____

_____

Flame Tests *continued*

Conclusions

1. Inferring Conclusions What metal ions are in the unknown solution?

_____

_____

_____

2. Evaluating Methods How would you characterize the flame test with respect to its sensitivity? What difficulties could there be when identifying ions by the flame test?

_____

_____

_____

_____

_____

EXTENSIONS

1. Inferring Conclusions A student performed flame tests on several unknowns and observed that they all were shades of red. What should the student do to correctly identify these substances? Explain your answer.

_____

_____

_____

2. Applying Ideas During a flood, the labels from three bottles of chemicals were lost. The three unlabeled bottles of white solids were known to contain the following: strontium nitrate, ammonium carbonate, and potassium sulfate. Explain how you could easily test the substances and relabel the three bottles. (Hint: Ammonium ions do not provide a distinctive flame color.)

_____

_____

_____

_____

Quick Lab

Designing Your Own Periodic Table

Can you design your own periodic table using information similar to that available to Mendeleev?

MATERIALS

• index cards

PROCEDURE

1. Write down the information available for each element on separate index cards. The following information is appropriate: a letter of the alphabet (A, B, C, etc.) to identify each element; atomic mass; state; density; melting point; boiling point; and any other readily observable physical properties. Do not write the name of the element on the index card, but keep a separate list indicating the letters you have assigned to each element.

2. Organize the cards for the elements in a logical pattern as you think Mendeleev might have done.

DISCUSSION

1. Keeping in mind that the information you have is similar to that available to Mendeleev in 1869, answer the following questions.

   a. Why are atomic masses given instead of atomic numbers?

   _____

   _____

   b. Can you identify each element by name?

   _____

   _____

2. How many groups of elements, or families, are in your periodic table? How many periods, or series, are in the table?

   _____

   _____

3. Predict the characteristics of any missing elements. When you have finished, check your work using your separate list of elements and a periodic table.

   _____

   _____

(Inquiry)

The Mendeleev Lab of 1869

Russian chemist Dmitri Mendeleev is generally credited with being the first chemist to observe that patterns emerge when the elements are arranged according to their properties. Mendeleev's arrangement of the elements was unique because he left blank spaces for elements that he claimed were undiscovered as of 1869. Mendeleev was so confident that he even predicted the properties of these undiscovered elements. His predictions were eventually proven to be quite accurate, and these new elements fill the spaces that originally were blank in his table. Use your knowledge of the periodic table to determine the identity of each of the nine unknown elements in this activity. These unknown elements are from the periodic table's groups that are listed below. Each of these groups contains at least one unknown element.

1   2   11   13   14   17   18

None of the known elements serves as one of the nine unknown elements. No radioactive elements are used during this experiment. The relevant radioactive elements include Fr, Ra, At, and Rn. You may not use your textbook or other reference materials. You have been provided with enough information to determine each of the unknown elements.

OBJECTIVES

- **Observe** the physical properties of common elements.

- **Observe** the properties and trends in the elements on the periodic table.

- **Draw conclusions** and **identify** unknown elements based on observed trends in properties.

MATERIALS

- blank periodic table

- elemental samples: Ar, C, Sn, and Pb

- note cards, 3 in. × 5 in.

- periodic table with element names and symbols, period numbers, and group numbers

- element cards

Always wear safety goggles and a lab apron to protect your eyes and clothing. If you get a chemical in your eyes, immediately flush the chemical out at the eyewash station while calling to your teacher. Know the locations of the emergency lab shower and the eyewash station and the procedures for using them.

The Mendeleev Lab of 1869 *continued*

PREPARATION

1. Use the **Data Table** provided to record the properties of each unknown that you test.

2. Use the note cards to copy the information listed on each of the sample cards. If the word *observe* is listed, you will need to visually inspect the sample and then write the observation in the appropriate space.

PROCEDURE

1. Arrange the note cards of the known elements in a rough representation of the periodic table. In other words, all of the known elements from Group 1 should be arranged in the appropriate order. Arrange all of the other cards accordingly.

2. Inspect the properties of the unknowns to see where properties would best "fit" the trends of the elements of each group.

3. Assign the proper element name to each of the unknowns. Add the symbol for each one of the unknown elements to your data table.

DISPOSAL

4. Clean up your lab station, and return the leftover note cards and samples of the elements to your teacher. Do not pour any of the samples down the drain or place them in the trash unless your teacher directs you to do so. Wash your hands thoroughly before you leave the lab and after all your work is finished.

Data Table		
Unknown	**Properties**	**Element**

Analysis

1. **Organizing Ideas** In what order did your group arrange the properties to determine the unknowns? Explain your reasoning. Would a different order have been better? If so, what is the better order and why?

_____

_____

_____

_____

_____

_____

_____

_____

_____

2. **Evaluating Methods** What properties were the most useful in sorting the unknowns? What properties were the least useful? Explain your answer.

_____

_____

_____

_____

_____

_____

_____

_____

_____

Conclusions

1. **Interpreting Information** Summarize your group's reasoning for the assignment of each unknown. Explain in a few sentences exactly how you predicted the identity of the nine unknown elements.

_____

_____

_____

_____

_____

_____

_____

_____

EXTENSIONS

1. **Predicting Outcomes** Use only the data from your group's experiment to predict the properties of the not yet discovered element, which has an atomic number of 120 (assuming it does not radioactively decay).

_____

_____

_____

_____

_____

_____

_____

_____

Inquiry

Types of Bonding in Solids

The purpose of this experiment is to relate certain properties of solids to the type of bonding the solids have. These observable properties depend on the type of bonding that holds the molecules, atoms, or ions together in each solid. Depending on the type of bonding, solids may be described as ionic, molecular, metallic, or covalent network solids. The properties to be studied are relative melting point, solubility in aqueous solution, and electrical conductivity.

OBJECTIVES

• **Observe** the physical properties of different solids.

• **Relate** knowledge of these properties to the type of bonding in each solid.

• **Identify** the type of bonding in an unknown solid.

MATERIALS

• beakers, 50 mL (6)

• Bunsen burner

• copper wire

• deionized water

• evaporating dishes or crucibles (6)

• graduated cylinder, 10 mL

• aluminum shot

• LED conductivity tester

• silicon dioxide (sand)

• sodium chloride (NaCl)

• spatula

• sucrose

• test tubes, small, with solid rubber stoppers (6)

• test-tube rack

• tongs

• unknown substance

• wire gauze, support stand, iron ring, and clay triangle

Always wear safety goggles and a lab apron to protect your eyes and clothing. If you get a chemical in your eyes, immediately flush the chemical out at the eyewash station while calling to your teacher. Know the locations of the emergency lab shower and the eyewash station and the procedures for using them.

Do not touch any chemicals. If you get a chemical on your skin or clothing, wash the chemical off at the sink while calling to your teacher. Make sure you carefully read the labels and follow the precautions on all containers of chemicals that you use. If there are no precautions stated on the label, ask your teacher what precautions you should follow. Do not taste any chemicals or items used in the laboratory. Never return leftovers to their original container; take only small amounts to avoid wasting supplies.

Types of Bonding in Solids *continued*

 When using a Bunsen burner, confine long hair and loose clothing. If your clothing catches on fire, WALK to the emergency lab shower and use it to put out the fire. When heating a substance in a test tube, the mouth of the test tube should point away from where you and others are standing. Watch the test tube at all times to prevent the contents from boiling over.

Never put broken glass in a regular waste container. Broken glass should be disposed of separately according to your teacher's instructions.

PREPARATION

Use the **Data Table** to record the results of melting, water solubility, solid conductivity, aqueous solution conductivity, and type of bonding of each substance tested.

PROCEDURE

1. Place 1 g samples of each substance into separate evaporating dishes.

2. Touch the electrodes of the conductivity tester to each solid. After each test, rinse with deionized water and carefully dry the electrodes. Note which substances conducted electricity.

3. Place one evaporating dish on a triangle, and heat with a Bunsen burner. As soon as a solid melts, remove the flame.

4. Repeat this procedure for every substance. Do not heat any substance for more than 5 min. There may be some substances that will not melt.

5. Note which substances melted and how long the substances took to melt.

6. Place five test tubes in the test-tube rack. Place 0.5 g of each solid into its own individual test tube. Add 5 mL of deionized water to each test tube. Stopper and shake each test tube in an attempt to dissolve the solid.

7. Note which substances dissolved in the water.

8. Place the solutions or mixtures into separate 50 mL beakers, and immerse the electrodes of the conductivity tester. Rinse the electrodes with the solvent (deionized water) before and after each test. Note which substances conduct electricity.

DISPOSAL

9. Dispose of solids and solutions in containers designated by your teacher.

10. Clean all equipment and return it to its proper place.

11. Wash your hands thoroughly after cleaning up your area and equipment.

Types of Bonding in Solids *continued*

Data Table						
	NaCl	**Sucrose**	**Copper wire**	**Iron filings**	**SiO₂**	**CaCl₂**
Melting						
Water solubility						
Solid conductivity						
Aqueous solution (or mixture) conductivity						
Type of bonding						

Analysis

1. Analyzing Methods Why did you rinse the electrodes before each conductivity test?

_____

_____

_____

_____

_____

2. Analyzing Methods Why did you use deionized water in making the solutions?

_____

_____

_____

_____

_____

_____

Types of Bonding in Solids *continued*

3. Organizing Data List the results that each type of bonding should show.

_____

_____

_____

_____

_____

_____

Conclusions

1. Inferring Conclusions What type of bonding describes each substance? Explain your reasoning.

_____

_____

_____

_____

_____

_____

_____

_____

_____

_____

_____

_____

_____

_____

_____

_____

_____

Types of Bonding in Solids *continued*

2. Inferring Conclusions Comparing the properties of your unknown solid with the properties of the known solids, determine the type of bonding present in your unknown solid.

_____

_____

_____

EXTENSIONS

1. Evaluating Methods Is it possible, for a specific type of bonding, for these properties to vary from what was observed in this experiment? If so, give an example of such a variance.

_____

_____

_____

_____

_____

_____

_____

2. Applying Conclusions Think about diamond. What would you predict to be the results of this experiment performed on diamond, and what would you predict the bond type to be?

_____

_____

_____

_____

_____

Skills Practice

Determining the Empirical Formula of Magnesium Oxide

This gravimetric analysis involves the combustion of magnesium metal in air to synthesize magnesium oxide. The mass of the product is greater than the mass of magnesium used because oxygen reacts with the magnesium metal. As in all gravimetric analyses, success depends on attaining a product yield near 100%. Therefore, the product will be heated and cooled and have its mass measured until two of these mass measurements are within 0.02% of one another. When the masses of the reactant and product have been carefully measured, the amount of oxygen used in the reaction can be calculated. The ratio of oxygen to magnesium can then be established, and the empirical formula of magnesium oxide can be determined.

OBJECTIVES

- **Measure** the mass of magnesium oxide.

- **Perform** a synthesis reaction by using gravimetric techniques.

- **Determine** the empirical formula of magnesium oxide.

- **Calculate** the class average and standard deviation for moles of oxygen used.

MATERIALS

- 15 cm magnesium ribbon, 2
- 25 mL beaker
- Bunsen burner assembly
- clay triangle
- crucible and lid, metal or ceramic
- crucible tongs
- distilled water
- eyedropper or micropipet
- ring stand

Always wear safety goggles and a lab apron to protect your eyes and clothing. If you get a chemical in your eyes, immediately flush the chemical out at the eyewash station while calling to your teacher. Know the locations of the emergency lab shower and the eyewash station and the procedures for using them.

When using a Bunsen burner, confine long hair and loose clothing. If your clothing catches on fire, WALK to the emergency lab shower and use it to put out the fire. When heating a substance in a test tube, the mouth of the test tube should point away from where you and others are standing. Watch the test tube at all times to prevent the contents from boiling over.

Never put broken glass in a regular waste container. Broken glass should be disposed of separately according to your teacher's instructions.

PREPARATION

Use the **Data Table** provided to record your data.

Determining the Empirical Formula of Magnesium Oxide *continued*

PROCEDURE

1. Construct a setup for heating a crucible as demonstrated in the Pre-Laboratory Procedure "Gravimetric Analysis."

2. Heat the crucible and lid for 5 min to burn off any impurities.

3. Cool the crucible and lid to room temperature. Measure their combined mass, and record the measurement on line 3 of the **Data Table.**

   NOTE: Handle the crucible and lid with crucible tongs at all times during this experiment. Such handling prevents burns and the transfer of dirt and oil from your hands to the crucible and lid.

4. Polish a 15 cm strip of magnesium with steel wool. The magnesium should be shiny. Cut the strip into small pieces to make the reaction proceed faster, and place the pieces in the crucible.

5. Cover the crucible with the lid, and measure the mass of the crucible, lid, and metal. Record the measurement on line 1 of the **Data Table.**

6. Use tongs to replace the crucible on the clay triangle. Heat the covered crucible gently. Lift the lid occasionally to allow air in.

   CAUTION: Do not look directly at the burning magnesium metal. The brightness of the light can blind you.

7. When the magnesium appears to be fully reacted, partially remove the crucible lid and continue heating for 1 min.

8. Remove the burner from under the crucible. After the crucible has cooled, use an eyedropper to carefully add a few drops of water to decompose any nitrides that may have formed.

   CAUTION: Use care when adding water. Using too much water can cause the crucible to crack.

9. Cover the crucible completely. Replace the burner under the crucible, and continue heating for about 30 to 60 s.

10. Turn off the burner. Cool the crucible, lid, and contents to room temperature. Measure the mass of the crucible, lid, and product. Record the measurement in the margin of the **Data Table.**

11. Replace the crucible, lid, and contents on the clay triangle, and reheat for another 2 min. Cool to room temperature, and remeasure the mass of the crucible, lid, and contents. Compare this mass measurement with the measurement obtained in step **10.** If the new mass is ±0.02% of the mass in step **10,** record the new mass on line 2 of your data table and go on to step **12.** If not, your reaction is still incomplete, and you should repeat step **11.**

12. Clean the crucible, and repeat steps **2–11** with a second strip of magnesium ribbon. Record your measurements under Trial 2 in the **Data Table.**

Determining the Empirical Formula of Magnesium Oxide *continued*

DISPOSAL

13. Put the solid magnesium oxide in the designated waste container. Return any unused magnesium ribbon to your teacher. Clean your equipment and lab station. Thoroughly wash your hands after completing the lab session and cleanup.

Data Table		
	Trial 1	**Trial 2**
1. Mass of crucible, lid, and metal (g)		
2. Mass of crucible, lid, and product (g)		
3. Mass of crucible and lid (g)		

Analysis

1. Applying Ideas Calculate the mass of the magnesium metal and the mass of the product.

2. Evaluating Data Determine the mass of the oxygen consumed.

3. Applying Ideas Calculate the number of moles of magnesium and the number of moles of oxygen in the product.

| Determining the Empirical Formula of Magnesium Oxide *continued*

Conclusions

1. Inferring Relationships Determine the empirical formula for magnesium
oxide, Mg_xO_y.

_____

_____

_____

_____

(Quick Lab)

Balancing Equations Using Models

How can molecular models and formula-unit ionic models be used to balance chemical equations and classify chemical reactions?

MATERIALS

- large and small gumdrops in at least four different colors
- toothpicks

Always wear safety goggles and a lab apron to protect your eyes and clothing. If you get a chemical in your eyes, immediately flush the chemical out at the eyewash station while calling to your teacher. Know the locations of the emergency lab shower and the eyewash station and the procedures for using them.

PROCEDURE

Examine the partial equations in Groups A–E. Using different-colored gumdrops to represent atoms of different elements, make models of the reactions by connecting the appropriate "atoms" with toothpicks. Use your models to (1) balance equations (a) and (b) in each group, (2) determine the products for reaction (c) in each group, and (3) complete and balance each equation (c). Finally, (4) classify each group of reactions by type.

Group A

a. $H_2 + Cl_2 \longrightarrow HCl$

_____

b. $Mg + O_2 \longrightarrow MgO$

_____

c. $BaO + H_2O \longrightarrow$ _____

_____

reaction type: _____

Group B

a. $H_2CO_3 \longrightarrow CO_2 + H_2O$

_____

b. $KClO_3 \longrightarrow KCl + O_2$

_____

c. $H_2O \xrightarrow{\text{electricity}}$ _____

_____

reaction type: _____

Balancing Equations Using Models *continued*

Group C

   a. $Ca + H_2O \longrightarrow Ca(OH)_2 + H_2$

   _____

   b. $KI + Br_2 \longrightarrow KBr + I_2$

   _____

   c. $Zn + HCl \longrightarrow$ _____

   _____

reaction type: _____

Group D

   a. $AgNO_3 + NaCl \longrightarrow AgCl + NaNO_3$

   _____

   b. $FeS + HCl \longrightarrow FeCl_2 + H_2S$

   _____

   c. $H_2SO_4 + KOH \longrightarrow$ _____

   _____

reaction type: _____

Group E

   a. $CH_4 + O_2 \longrightarrow CO_2 + H_2O$

   _____

   b. $CO + O_2 \longrightarrow CO_2$

   _____

   c. $C_3H_8 + O_2 \longrightarrow$ _____

   _____

reaction type: _____

Inquiry)   **DATASHEET FOR IN-TEXT LAB**

Blueprint Paper

Blueprint paper is prepared by coating paper with a solution of two soluble iron(III) salts—potassium hexacyanoferrate(III), commonly called potassium ferricyanide, and iron(III) ammonium citrate. These two salts do not react with each other in the dark. However, when exposed to UV light, the iron(III) ammonium citrate is converted to an iron(II) salt. Potassium hexacyanoferrate(III), $K_3Fe(CN)_6$, reacts with iron(II) ion, Fe^{2+}, to produce an insoluble blue compound, $KFeFe(CN)_6 \cdot H_2O$. In this compound, iron appears to exist in both the $+2$ and $+3$ oxidation states.

A blueprint is made by using black ink to make a sketch on a piece of tracing paper or clear, colorless plastic. This sketch is placed on top of a piece of blueprint paper and exposed to ultraviolet light. Wherever the light strikes the paper, the paper turns blue. The paper is then washed to remove the soluble unexposed chemical and is allowed to dry. The result is a blueprint—a blue sheet of paper with white lines.

OBJECTIVE

• **Prepare** blueprint paper and create a blueprint.

MATERIALS

• 10% iron(III) ammonium citrate solution

• 10% potassium hexacyanoferrate(III) solution

• 25 mL graduated cylinders, 2

• corrugated cardboard, 20 cm × 30 cm, 2 pieces

• glass stirring rod

• Petri dish

• thumbtacks, 4

• tongs

• white paper, 8 cm × 15 cm, 1 piece

Always wear safety goggles and a lab apron to protect your eyes and clothing. If you get a chemical in your eyes, immediately flush the chemical out at the eyewash station while calling to your teacher. Know the locations of the emergency lab shower and the eyewash station and the procedures for using them.

Do not touch any chemicals. If you get a chemical on your skin or clothing, wash the chemical off at the sink while calling to your teacher. Make sure you carefully read the labels and follow the precautions on all containers of chemicals that you use. If there are no precautions stated on the label, ask your teacher what precautions you should follow. Do not taste any chemicals or items used in the laboratory. Never return leftovers to their original container; take only small amounts to avoid wasting supplies.

Blueprint Paper *continued*

PROCEDURE

1. Pour 15 mL of a 10% solution of potassium hexacyanoferrate(III) solution into a Petri dish. With most of the classroom lights off or dimmed, add 15 mL of 10% iron(III) ammonium citrate solution. Stir the mixture.

2. Write your name on an 8 cm × 15 cm piece of white paper. Carefully coat one side of the piece of paper by using tongs to drag it over the top of the solution in the Petri dish.

3. With the coated side up, tack your wet paper to a piece of corrugated cardboard, and cover the paper with another piece of cardboard. **Wash your hands before proceeding to step 4.**

4. Take your paper and cardboard assembly outside into the direct sunlight. Remove the top piece of cardboard so that the paper is exposed. Quickly place an object such as a fern, a leaf, or a key on the paper. If it is windy, you may need to put small weights, such as coins, on the object to keep it in place.

5. After about 20 min, remove the object and again cover the paper with the cardboard. Return to the lab, remove the tacks, and thoroughly rinse the blueprint paper under cold running water. Allow the paper to dry. In your notebook, record the amount of time that the paper was exposed to sunlight.

DISPOSAL

6. Clean all equipment and your lab station. Return equipment to its proper place. Dispose of chemicals and solutions in the containers designated by your teacher. Do not pour any chemicals down the drain or in the trash unless your teacher directs you to do so. Wash your hands thoroughly before you leave the lab and after all work is finished.

Analysis

1. **Relating Ideas** Why is the iron(III) ammonium citrate solution stored in a brown bottle?

_____

_____

_____

_____

2. **Organizing Ideas** When iron(III) ammonium citrate is exposed to light, the oxidation state of the iron changes. What is the new oxidation state of the iron?

_____

Blueprint Paper *continued*

3. Analyzing Methods What substances were washed away when you rinsed the blueprint in water after it had been exposed to sunlight? (Hint: Compare the solubilities of the two ammonium salts that you used to coat the paper and of the blue product that formed.)

_____

_____

_____

_____

_____

_____

_____

_____

Conclusions

1. Applying Ideas Insufficient washing of the exposed blueprints results in a slow deterioration of images. Suggest a reason for this deterioration.

_____

_____

_____

_____

2. Relating Ideas Photographic paper can be safely exposed to red light in a darkroom. Do you think the same would be true of blueprint paper? Explain your answer.

_____

_____

_____

_____

_____

Blueprint Paper *continued*

EXTENSIONS

1. **Applying Ideas** How could you use this blueprint paper to test the effectiveness of a brand of sunscreen lotion?

_____

_____

_____

_____

2. **Designing Experiments** Can you think of ways to improve this procedure? If so, ask your teacher to approve your plan, and create a new blueprint. Evaluate both the efficiency of the procedure and the quality of blueprint.

_____

_____

_____

_____

_____

_____

_____

_____

DATASHEET FOR IN-TEXT LAB

Limiting Reactants in a Recipe

MATERIALS

- 1/2 cup sugar
- 1/2 cup brown sugar
- 1 1/3 stick margarine (at room temperature)
- 1 egg
- 1/2 tsp. salt
- 1 tsp. vanilla
- 1/2 tsp. baking soda

- 1 1/2 cup flour
- 1 1/3 cup chocolate chips
- mixing bowl
- mixing spoon
- measuring spoons and cups
- cookie sheet
- oven preheated to 350°F

PROCEDURE

1. In the mixing bowl, combine the sugars and margarine together until smooth. (An electric mixer will make this process go much faster.)
2. Add the egg, salt, and vanilla. Mix well.
3. Stir in the baking soda, flour, and chocolate chips. Chill the dough for an hour in the refrigerator for best results.
4. Divide the dough into 24 small balls about 3 cm in diameter. Place the balls on an ungreased cookie sheet.
5. Bake at 350°F for about 10 minutes, or until the cookies are light brown.

   Yield: 24 cookies

DISCUSSION

1. Suppose you are given the following amounts of ingredients:

   1 dozen eggs

   24 tsp. of vanilla

   1 lb. (82 tsp.) of salt

   1 lb. (84 tsp.) of baking soda

   3 cups of chocolate chips

   5 lb. (11 cups) of sugar

   2 lb. (4 cups) of brown sugar

   1 lb. (4 sticks) of margarine

Limiting Reactants in a Recipe *continued*

a. For each ingredient, calculate how many cookies could be prepared if all of that ingredient were consumed. (For example, the recipe shows that using 1 egg—with the right amounts of the other ingredients—yields 24 cookies. How many cookies can you make if the recipe is increased proportionately for 12 eggs?)

_____

_____

_____

_____

_____

_____

_____

_____

b. To determine the limiting reactant for the new ingredients list, identify which ingredient will result in the fewest number of cookies.

_____

_____

c. What is the maximum number of cookies that can be produced from the new amounts of ingredients?

_____

_____

Skills Practice

Stoichiometry and Gravimetric Analysis

This gravimetric analysis involves a double-displacement reaction between strontium chloride, $SrCl_2$, and sodium carbonate, Na_2CO_3. This type of reaction can be used to determine the amount of a carbonate compound in a solution. For accurate results, essentially all of the reactant of unknown amount must be converted into product. If the mass of the product is carefully measured, you can use stoichiometric calculations to determine how much of the reactant of unknown amount was involved in the reaction.

OBJECTIVES

- **Observe** the double-displacement reaction between solutions of strontium chloride and sodium carbonate.

- **Demonstrate** proficiency with gravimetric methods.

- **Measure** the mass of the precipitate that forms.

- **Relate** the mass of the precipitate that forms to the mass of the reactants before the reaction.

- **Calculate** the mass of sodium carbonate in a solution of unknown concentration.

MATERIALS

- 15 mL Na_2CO_3 solution of unknown concentration
- 50 mL 0.30 M $SrCl_2$ solution
- 50 mL graduated cylinder
- 250 mL beakers, 2
- balance
- beaker tongs
- distilled water
- drying oven
- filter paper
- glass funnel or Büchner funnel with related equipment
- glass stirring rod
- paper towels
- ring and ring stand
- spatula
- water bottle

![safety goggles and lab apron icons] **Always wear safety goggles and a lab apron to protect your eyes and clothing.** If you get a chemical in your eyes, immediately flush the chemical out at the eyewash station while calling to your teacher. Know the locations of the emergency lab shower and the eyewash station and the procedures for using them.

![hazard icon] **Do not touch any chemicals.** If you get a chemical on your skin or clothing, wash the chemical off at the sink while calling to your teacher. Make sure you carefully read the labels and follow the precautions on all containers of chemicals that you use. If there are no precautions stated on the label, ask your teacher what precautions you should follow. Do not taste any chemicals or items used in the

laboratory. Never return leftovers to their original container; take only small amounts to avoid wasting supplies.

 Never put broken glass in a regular waste container. Broken glass should be disposed of separately according to your teacher's instructions.

PREPARATION

1. Use the data table provided to record your data.

2. Clean all of the necessary lab equipment with soap and water, and rinse with distilled water.

3. Measure the mass of a piece of filter paper to the nearest 0.01 g, and record it in your table.

4. Set up a filtering apparatus. Use the Pre-Laboratory Procedure "Extraction and Filtration."

5. Label a paper towel with your name and the date. Place the towel in a clean, dry 250 mL beaker, and measure and record the mass of the paper towel and beaker to the nearest 0.01 g.

PROCEDURE

1. Measure about 15 mL of the Na_2CO_3 solution into the graduated cylinder. Record this volume to the nearest 0.5 mL. Pour the Na_2CO_3 solution into an empty 250 mL beaker. Carefully wash the graduated cylinder, and rinse it with distilled water.

2. Measure about 25 mL of the 0.30 M $SrCl_2$ solution into the graduated cylinder. Record this volume to the nearest 0.5 mL. Pour the $SrCl_2$ solution into the beaker with the Na_2CO_3 solution. Gently stir with a glass stirring rod.

3. Measure another 10 mL of the $SrCl_2$ solution into the graduated cylinder. Record the volume to the nearest 0.5 mL. Slowly add the solution to the beaker, and stir gently. Repeat this step until no more precipitate forms.

4. Slowly pour the mixture into the funnel. Do not overfill the funnel—some of the precipitate could be lost between the filter paper and the funnel.

5. Rinse the beaker several more times with distilled water. Pour the rinse water into the funnel each time.

6. After all of the solution and rinses have drained through the funnel, use distilled water to slowly rinse the precipitate on the filter paper in the funnel to remove any soluble impurities.

7. Carefully remove the filter paper from the funnel, and place it on the paper towel that you labeled with your name. Unfold the filter paper, and place the paper towel, filter paper, and precipitate in the rinsed beaker. Then, place the beaker in the drying oven. For best results, allow the precipitate to dry overnight.

Stoichiometry and Gravimetric Analysis *continued*

8. Using beaker tongs, remove your sample from the oven, and let it cool. Record the total mass of the beaker, paper towel, filter paper, and precipitate to the nearest 0.01 g.

DISPOSAL

9. Dispose of the precipitate and the filtrate in designated waste containers. Clean up all equipment after use, and dispose of substances according to your teacher's instructions. Wash your hands thoroughly after all lab work is finished.

Data Table	
Volume of Na_2CO_3 solution added	
Volume of $SrCl_2$ solution added	
Mass of dry filter paper	
Mass of beaker with paper towel	
Mass of beaker with paper towel, filter paper, and precipitate	

Analysis

1. Organizing Ideas Write a balanced equation for the reaction. What is the precipitate?

_____

_____

_____

2. Applying Ideas Calculate the mass of the dry precipitate. Calculate the number of moles of precipitate produced in the reaction.

3. Applying Ideas How many moles of Na_2CO_3 were present in the 15 mL sample? How many grams of Na_2CO_3 were present?

Conclusions

1. Applying Conclusions There are 0.30 mol $SrCl_2$ in every liter of solution. Calculate the number of moles of $SrCl_2$ that were added. What is the limiting reactant?

"Wet" Dry Ice

The phase diagram for carbon dioxide shows that CO_2 can exist only as a gas at ordinary room temperature and pressure. To observe the transition of solid CO_2 to liquid CO_2, you must increase the pressure until it is at or above the triple point pressure.

OBJECTIVES

• **Interpret** a phase diagram.

• **Observe** the melting of CO_2 while varying pressure.

• **Relate** observations of CO_2 to its phase diagram.

MATERIALS

• 4–5 g CO_2 as dry ice, broken into rice-sized pieces

• forceps

• metric ruler

• plastic pipets, 5 mL, shatterproof

• pliers

• scissors

• transparent plastic cup

Always wear safety goggles and a lab apron to protect your eyes and clothing. If you get a chemical in your eyes, immediately flush the chemical out at the eyewash station while calling to your teacher. Know the locations of the emergency lab shower and the eyewash station and the procedures for using them.

Do not touch any chemicals. If you get a chemical on your skin or clothing, wash the chemical off at the sink while calling to your teacher. Make sure you carefully read the labels and follow the precautions on all containers of chemicals that you use. If there are no precautions stated on the label, ask your teacher what precautions you should follow. Do not taste any chemicals or items used in the laboratory. Never return leftovers to their original container; take only small amounts to avoid wasting supplies.

PROCEDURE

1. Use forceps to place 2–3 very small pieces of dry ice on the table, and observe them until they have completely sublimed. Record your observations in the space provided. **Caution:** Dry ice will freeze skin very quickly. Do not attempt to pick up the dry ice with your fingers.

2. Fill a plastic cup with tap water to a depth of 4–5 cm.

3. Cut the tapered end (tip) off the graduated pipet.

4. Use forceps to carefully slide 8–10 pieces of dry ice down the stem and into the bulb of the pipet.

5. Use a pair of pliers to clamp the opening of the pipet stem securely shut so that no gas can escape. Use the pliers to hold the tube and to lower the pipet into the cup just until the bulb is submerged. From the side of the cup, observe the behavior of the dry ice. Record your observations in the space provided.

6. As soon as the dry ice has begun to melt, quickly loosen the pliers while still holding the bulb in the water. Observe the CO_2 and record your observations in the space provided.

7. Tighten the pliers again, and record your observations.

8. Repeat Procedure steps 6 and 7 as many times as possible.

DISPOSAL

9. Clean all apparatus and your lab station. Return equipment to its proper place. Dispose of chemicals and solutions in the containers designated by your teacher. Do not pour any chemicals down the drain or place them in the trash unless your teacher directs you to do so. Wash your hands thoroughly before you leave the lab and after all work is finished.

OBSERVATIONS

Procedure 1 _____

_____

_____

_____

Procedure 5 _____

_____

_____

_____

Procedure 6 and 7 _____

_____

_____

_____

| "Wet" Dry Ice *continued*

Analysis

1. Analyzing Results What differences did you observe between the subliming and the melting of CO_2?

_____

_____

_____

2. Analyzing Methods As you melted the CO_2 sample over and over, why did it eventually disappear? What could you have done to make the sample last longer?

_____

_____

_____

3. Analyzing Methods What purpose(s) do you suppose the water in the cup served?

_____

_____

_____

_____

_____

_____

_____

_____

_____

_____

_____

_____

_____

"Wet" Dry Ice *continued*

EXTENSIONS

1. **Predicting Outcomes** What would have happened if fewer pieces of dry ice (only 1 or 2) had been placed inside the pipet bulb? If time permits, test your prediction.

_____

_____

_____

_____

_____

2. **Predicting Outcomes** What might have happened if too much dry ice (20 or 30 pieces, or example) had been placed inside the pipet bulb? How quickly would the process have occurred? If time permits, test your prediction.

_____

_____

_____

_____

_____

_____

3. **Predicting Outcomes** What would have happened if the pliers had not been released once the dry ice melted? If time permits, test your prediction.

_____

_____

_____

_____

Name _____ Class _____ Date _____

Do different gases diffuse at different rates?

MATERIALS

- household ammonia
- perfume or cologne
- two 250 mL beakers

- two watch glasses
- 10 mL graduated cylinder
- clock or watch with second hand

Always wear safety goggles and a lab apron to protect your eyes and clothing. If you get a chemical in your eyes, immediately flush the chemical out at the eyewash station while calling to your teacher. Know the locations of the emergency lab shower and the eyewash station and the procedures for using them.

Do not touch any chemicals. If you get a chemical on your skin or clothing, wash the chemical off at the sink while calling to your teacher. Make sure you carefully read the labels and follow the precautions on all containers of chemicals that you use. If there are no precautions stated on the label, ask your teacher what precautions you should follow. Do not taste any chemicals or items used in the laboratory. Never return leftovers to their original container; take only small amounts to avoid wasting supplies.

PROCEDURE

Record all of your results in the **Data Table.**

1. Outdoors or in a room separate from the one in which you will carry out the rest of the investigation, pour approximately 10 mL of the household ammonia into one of the 250 mL beakers, and cover it with a watch glass. Pour roughly the same amount of perfume or cologne into the second beaker. Cover it with a watch glass also.

2. Take the two samples you just prepared into a large, draft-free room. Place the samples about 12 to 15 ft apart and at the same height. Position someone as the observer midway between the two beakers. Remove both watch glass covers at the same time.

3. Note whether the observer smells the ammonia or the perfume first. Record how long this takes. Also, record how long it takes the vapor of the other substance to reach the observer. Air the room after you have finished.

DISPOSAL

4. Check with your teacher for the proper disposal procedures. Always wash your hands thoroughly after cleaning up the lab area and equipment.

❘ Diffusion *continued*

Data Table

	Ammonia	**Perfume/cologne**
Location		
Distance		
Time (s)		

DISCUSSION

1. What do the times that the two vapors took to reach the observer show about the two gases?

_____

_____

2. What factors other than molecular mass (which determines diffusion rate) could affect how quickly the observer smells each vapor?

_____

_____

_____

Microscale

Mass and Density of Air at Different Pressures

You have learned that the amount of gas present, the volume of the gas, the temperature of the gas sample, and the gas pressure are related to one another. If the volume and temperature of a gas sample are held constant, the mass of the gas and the pressure that the gas exerts are related in a simple way.

In this investigation, you will use an automobile tire pressure gauge to measure the mass of a bottle and the air that the bottle contains for several air pressures. A tire pressure gauge measures "gauge pressure," meaning the added pressure in the tire in addition to normal atmospheric air pressure. Gauge pressure is often expressed in the units pounds per square inch, gauge (psig) to distinguish them from absolute pressures in pounds per square inch (psi). You will graph the mass of the bottle plus air against the gas pressure and observe what kind of plot results. Extrapolating this plot in the proper way will let you determine both the mass and the volume of the empty bottle. This information will also allow you to calculate the density of air at various pressures.

OBJECTIVES

- **Measure** the pressure exerted by a gas.

- **Measure** the mass of a gas sample at different pressures.

- **Graph** the relationship between the mass and pressure of a gas sample.

- **Calculate** the mass of an evacuated bottle.

- **Calculate** the volume of a bottle.

- **Calculate** the density of air at different pressures.

MATERIALS

- automobile tire valve
- balance, centigram
- barometer
- cloth towel

- plastic soda bottle (2 or 3 L) or other heavy plastic bottle
- tire pressure gauge

Always wear safety goggles and a lab apron to protect your eyes and clothing. If you get a chemical in your eyes, immediately flush the chemical out at the eyewash station while calling to your teacher. Know the locations of the emergency lab shower and the eyewash station and the procedures for using them.

PREPARATION

Use the **Data Table** provided to record your data.

| Mass and Density of Air at Different Pressures *continued*

PROCEDURE

Your teacher will provide you with a bottle. This bottle contains air under considerable pressure, so handle it carefully. Do not unscrew the cap of the bottle.

1. Use the tire pressure gauge to measure the gauge pressure of the air in the bottle, as accurately as you can read the gauge. It might be convenient for one student to hold the bottle securely, wrapped in a cloth towel, while another student makes the pressure measurement. Record this pressure in the **Data Table.**

2. Measure the mass of the bottle plus the air it contains, to the nearest 0.01 g. Record this mass in the **Data Table.**

3. With one student holding the wrapped bottle, depress the tire stem valve carefully to allow some air to escape from the bottle until the observed gauge pressure has decreased by 5 to 10 psig. Then, repeat the measurements in steps **1** and **2.**

4. Repeat the steps of releasing some pressure (step **3**) and then measuring gauge pressure (step **1**) and measuring the mass (step **2**) until no more air comes out.

5. Now repeat steps **1** and **2** one last time. The gauge pressure should be zero; if it is not, then you probably have not released enough air, and you should depress the valve for a longer time. You should have at least five measurements of gauge pressure and mass, including this final set.

6. Read the atmospheric pressure in the room from the barometer, and record the reading in the **Data Table.**

DISPOSAL

7. Return all equipment to its proper place. Wash your hands thoroughly before you leave the lab and after all work is finished.

Data Table

Gauge Pressure (psig)	Mass of Bottle + Air (g)	Corrected Gas Pressure (psi)	Mass of Air (g)	Density of Air (g/cm³)

Mass and Density of Air at Different Pressures *continued*

Analysis

1. Organizing Data Correct each gauge pressure in your data table to the actual gas pressure in psi, by adding the barometric pressure (in psi) to each measured gauge pressure. Enter these results in the column "Corrected gas pressure."

2. Analyzing Data Make a graph of your data. Plot corrected gas pressure on the x-axis and mass of bottle plus air on the y-axis. The x-axis should run from 10 psi to at least 80 psi. The y-axis scale should allow extrapolation to corrected gas pressure of zero.

Mass and Density of Air at Different Pressures *continued*

3. **Analyzing Data** If your graph is a straight line, write an equation for the line in the form $y = mx + b$. If the graph is not a straight line, explain why, and draw the straight line that comes closest to including all of your data points. Give the equation of this line.

4. **Interpreting Data** What is the mass of the empty bottle? (Hint: When no more air escapes from the bottle in steps **4** and **5,** the bottle is not empty; it still contains air at 1 atm.)

_____

_____

_____

_____

_____

_____

_____

5. **Analyzing Data** For each of your readings, calculate the mass of air in the bottle. Enter these masses in the **Data Table.**

Mass and Density of Air at Different Pressures *continued*

6. **Interpreting Data** The density of air at typical laboratory conditions is
   1.19 g/L. Find the volume of the bottle.

7. **Interpreting Data** Calculate the density of air at each pressure for which you
   made measurements. Enter these density values in your data table.

Conclusions

1. **Inferring Relationships** Based on your results in this experiment, state the
   relationship between the mass of a gas sample and the gas pressure. Be sure
   to include limitations (that is, the quantities that must be kept constant).

   _____

   _____

   _____

Mass and Density of Air at Different Pressures *continued*

2. **Interpreting Graphics** Using your graph from item 2 of Analysis, predict the mass of the bottle plus air at a gauge reading of 60.0 psig. Estimate the mass of the gas in the bottle at that pressure.

_____

_____

_____

_____

_____

_____

Quick Lab)   **DATASHEET FOR IN-TEXT LAB**

Observing Solutions, Suspensions, and Colloids

MATERIALS

- balance
- 7 beakers, 400 mL
- clay
- cooking oil
- flashlight
- gelatin, plain
- hot plate (to boil H_2O)
- red food coloring

- sodium borate ($Na_2B_4O_7 \cdot 10H_2O$)
- soluble starch
- stirring rod
- sucrose
- test-tubes, 7
- test-tube rack
- water

Always wear safety goggles and a lab apron to protect your eyes and clothing. If you get a chemical in your eyes, immediately flush the chemical out at the eyewash station while calling to your teacher. Know the locations of the emergency lab shower and the eyewash station and the procedures for using them.

PROCEDURE

1. Prepare seven mixtures, each containing 250 mL of water and one of the following substances.

   a. 12 g of sucrose

   b. 3 g of soluble starch

   c. 5 g of clay

   d. 2 mL of food coloring

   e. 2 g of sodium borate

   f. 50 mL of cooking oil

   g. 3 g of gelatin

   Making the gelatin mixture: Soften the gelatin in 65 mL of cold water, and then add 185 mL of boiling water.

2. Observe the seven mixtures and their characteristics. Record the appearance of each mixture after stirring.

3. Transfer to individual test tubes 10 mL of each mixture that does not separate after stirring. Shine a flashlight on each mixture in a dark room. Make note of the mixtures in which the path of the light beam is visible.

Modern Chemistry   **215**   Solutions

Observing Solutions, Suspensions, and Colloids *continued*

DISCUSSION

1. Using your observations, classify each mixture as a solution, suspension, or colloid.

_____

_____

_____

_____

2. What characteristics did you use to classify each mixture?

_____

_____

_____

Skills Practice

DATASHEET FOR IN-TEXT LAB

Separation of Pen Inks by Paper Chromatography

Paper Chromatography

Details on this technique can be found in the Pre-Laboratory Procedure "Paper Chromatography" on page 848.

Writing Inks

Most ballpoint pen inks are complex mixtures, containing pigments or dyes that can be separated by paper chromatography.

Black inks can contain three or more colors; the number of colors depends on the manufacturer. Each ink formulation has a characteristic pattern that uniquely identifies it.

In this experiment you will develop radial paper chromatograms for four black ballpoint pen inks, using water as solvent. You will then repeat this process using isopropanol as the solvent. You will then measure the distance traveled by each of the individual ink components and the distance traveled by the solvent front. Finally, you will use these measurements to calculate the R_f factor for each component.

OBJECTIVES

- **Demonstrate** proficiency in qualitatively separating mixtures using paper chromatography.

- **Determine** the R_f factor(s) for each component of each tested ink.

- **Explain** how the inks are separated by paper chromatography.

- **Observe** the separation of a mixture by the method of paper chromatography.

MATERIALS

- 12 cm circular chromatography paper or filter paper, 2

- distilled water

- 1 filter paper wick, 2 cm equilateral triangle

- isopropanol

- numbered pens, each with a different black ink, 4

- pencil

- petri dish with lid

- scissors

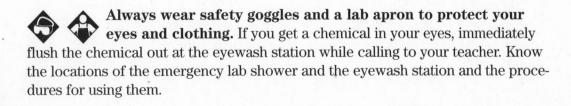 **Always wear safety goggles and a lab apron to protect your eyes and clothing.** If you get a chemical in your eyes, immediately flush the chemical out at the eyewash station while calling to your teacher. Know the locations of the emergency lab shower and the eyewash station and the procedures for using them.

| Separation of Pen Inks by Paper Chromatography *continued*

☠ **Do not touch any chemicals.** If you get a chemical on your skin or clothing, wash the chemical off at the sink while calling to your teacher. Make sure you carefully read the labels and follow the precautions on all containers of chemicals that you use. If there are no precautions stated on the label, ask your teacher what precautions you should follow. Do not taste any chemicals or items used in the laboratory. Never return leftovers to their original container; take only small amounts to avoid wasting supplies.

PREPARATION

1. Determine the formula, structure, polarity, density, and volatility at room temperature for water and isopropanol. The following titles are sources that provide general information on specific elements and compounds: *CRC Handbook of Chemistry and Physics*, *McGraw-Hill Dictionary of Chemical Terms*, and *Merck Index*.

2. Use the data tables provided to record your data.

PROCEDURE

Part A: Prepare a chromatogram using water as the solvent

1. Construct an apparatus for paper chromatography as described in the Pre-Laboratory Procedure on page 848. You will make only four dots. You will use ballpoint pens rather than micropipets to spot your paper.

2. After 15 min or when the water is about 1 cm from the outside edge of the paper, remove the paper from the Petri dish and allow the chromatogram to dry. Record in **Data Table 1** the colors that have separated from each of the four different black inks.

Part B: Prepare a chromatogram using isopropanol as the solvent

3. Repeat Procedure steps **1** to **2**, replacing the water in the Petri dish with isopropanol. Record in **Data Table 2** the colors that have separated from each of the four different black inks.

Part C: Determine R_f values for each component

4. After the chromatogram is dry, use a pencil to mark the point where the solvent front stopped.

5. With a ruler, measure the distance from the initial ink spot to your mark, and record this distance on the appropriate data table.

6. Make a small dot with your pencil in the center of each color band.

7. With a ruler, measure the distance from the initial ink spot to each dot separately, and record each distance on the appropriate data table.

Separation of Pen Inks by Paper Chromatography *continued*

8. Divide each value recorded in Procedure step **7** by the value recorded in Procedure step **5.** The result is the R_f value for that component. Record the R_f values in the appropriate data table. Tape or staple the chromatogram to the appropriate data table.

DISPOSAL

9. The water may be poured down the sink. Chromatograms and other pieces of filter paper may be discarded in the trash. The isopropanol solution should be placed in the waste disposal container designated by your teacher. Clean up your equipment and lab station. Thoroughly wash your hands after completing the lab session and cleanup.

DATA TABLE 1		Chromatogram Formed with Water							
Pen no.	Dot no.	Color 1		Color 2		Color 3		Color 4	
		Distance	R_f value	Distance	R_f value	Distance	R_f value	Distance	R_f value

DATA TABLE 2		Chromatogram Formed with Isopropanol							
Pen no.	Dot no.	Color 1		Color 2		Color 3		Color 4	
		Distance	R_f value	Distance	R_f value	Distance	R_f value	Distance	R_f value

Analysis

1. Evaluating Conclusions Is the color in each pen the result of a single dye or multiple dyes? Justify your answer.

_____

_____

_____

_____

_____

_____

_____

_____

_____

_____

Separation of Pen Inks by Paper Chromatography *continued*

2. Relating Ideas What can be said about the properties of a component ink that has an R_f value of 0.50?

_____

_____

_____

3. Analyzing Methods Suggest a reason for stopping the process when the solvent front is 1 cm from the edge of the filter paper rather than when it is even with the edge of the paper.

_____

_____

_____

4. Predicting Outcomes Predict the results of forgetting to remove the chromatogram from the water in the petri dish until the next day.

_____

_____

_____

_____

_____

Conclusions

1. Analyzing Results Compare the R_f values for the colors from pen number 2 when water was the solvent and the R_f values obtained when isopropanol was the solvent. Explain why they differ.

_____

_____

_____

2. Evaluating Methods Would you consider isopropanol a better choice for the solvent than water? Why or why not?

_____

_____

_____

Separation of Pen Inks by Paper Chromatography *continued*

3. Analyzing Conclusions Are the properties of the component that traveled the farthest in the water chromatogram likely to be similar to the properties of the component that traveled the farthest in the isopropanol chromatogram? Explain your reasoning.

_____

_____

_____

_____

_____

_____

4. Inferring Conclusions What can you conclude about the composition of the inks in ballpoint pens from your chromatogram?

_____

_____

_____

_____

_____

_____

_____

_____

_____

_____

_____

_____

_____

_____

_____

_____

Microscale

Testing Water for Ions

The physical and chemical properties of aqueous solutions are affected by small amounts of dissolved ions. For example, if a water sample has enough Mg^{2+} or Ca^{2+} ions, it does not create lather when soap is added. This is common in places where there are many minerals in the water (hard water). Other ions, such as Pb^{2+} and Co^{2+}, can accumulate in body tissues; therefore, solutions of these ions are poisonous.

Because some sources of water may contain harmful or unwanted substances, it is important to find out what ions are present. In this experiment, you will test various water samples for the presence of four ions: Fe^{3+}, Ca^{2+}, Cl^-, and SO_4^{2-}. Some of the samples may contain these ions in very small concentrations, so make very careful observations.

OBJECTIVES

• **Observe** chemical reactions involving aqueous solutions of ions.

• **Relate** observations of chemical properties to the presence of ions.

• **Infer** whether an ion is present in a water sample.

• **Apply** concepts concerning aqueous solutions of ions

MATERIALS

• 24-well microplate lid

• fine-tipped dropper bulbs, labeled, with solutions, 10

• overhead projector (optional)

• paper towels

• solution 1: reference (all ions)

• solution 2: distilled water (no ions)

• solution 3: tap water (may have ions)

• solution 4: bottled spring water (may have ions)

• solution 5: local river or lake water (may have ions)

• solution 6: solution X, prepared by your teacher (may have ions)

• solution A: NaSCN solution (test for Fe^{3+})

• solution B: $Na_2C_2O_4$ solution (test for Ca^{2+})

• solution C: $AgNO_3$ solution (test for Cl^-)

• solution D: $Sr(NO_3)_2$ solution (test for SO_4^{2-})

• white paper

⬧ ⬧ **Always wear safety goggles and a lab apron to protect your eyes and clothing.** If you get a chemical in your eyes, immediately flush the chemical out at the eyewash station while calling to your teacher. Know the locations of the emergency lab shower and the eyewash station and the procedures for using them.

Do not touch any chemicals. If you get a chemical on your skin or clothing, wash the chemical off at the sink while calling to your teacher. Make sure you carefully read the labels and follow the precautions on all containers of chemicals that you use. If there are no precautions stated on the label, ask your teacher what precautions you should follow. Do not taste any chemicals or items used in the laboratory. Never return leftovers to their original container; take only small amounts to avoid wasting supplies.

PREPARATION

1. Use the **Data Table** provided to record your observations.

2. Place the 24-well microplate lid in front of you on a white background. Label the columns and rows as instructed by your teacher. The coordinates will designate the individual circles. For example, the circle in the top right corner would be 1-D.

PROCEDURE

1. Obtain labeled dropper bulbs containing the six different solutions from your teacher.

2. Place a drop of the solution from bulb 1 into circles 1-A, 1-B, 1-C, and 1-D (the top row). Solution 1 contains all four of the dissolved ions, so these drops will show what a **positive** test for each ion looks like. **Be careful to keep the solutions in the appropriate circles. Any spills will cause poor results.**

3. Place a drop of the solution from bulb 2 into each of the circles in row 2. This solution is distilled water and should not contain any of the ions. It will show what a **negative** test looks like.

4. Place a drop from bulb 3 into each of the circles in row 3 and a drop from bulb 4 into each of the circles in row 4. Follow the same procedure for bulb 5 (into row 5) and bulb 6 (into row 6). These solutions may or may not contain ions. The materials list gives contents of each bulb.

5. Now that each circle contains a solution to be analyzed, use the solutions in bulbs A–D to test for the presence of the ions. Bulb A contains NaSCN, sodium thiocyanate, which reacts with any Fe^{3+} to form the complex ion $Fe(SCN)^{2+}$, which results in a deep red solution. Bulb B contains $Na_2C_2O_4$, sodium oxalate, which reacts with Ca^{2+} ions. Bulb C contains $AgNO_3$, silver nitrate, which reacts with Cl^- ions. Bulb D contains $Sr(NO_3)_2$, strontium nitrate, which reacts with SO_4^{2-} ions. The contents of bulbs B–D react with the specified ion to yield insoluble precipitates.

6. **Holding the tip of bulb A 1 to 2 cm above the drop of water to be tested,** add one drop of solution A to the drop of reference solution in circle 1-A and one drop to the distilled water in circle 2-A. Circle 1-A should show a positive test, and circle 2-A should show a negative test. In the **Data Table,** record your observations about what the positive and negative tests look like.

| Testing Water for Ions *continued*

7. Use the NaSCN solution in bulb A to test the rest of the water drops in column A to determine whether they contain the Fe^{3+} ion. Record your observations in the **Data Table.** For each of the tests in which the ion was present, specify whether it seemed to be at a high, moderate, or low concentration.

8. Follow the procedure used for bulb A with bulbs B, C, and D to test for the other ions. Record your observations about the test results. Specify whether the solutions contained Ca^{2+}, Cl^-, or SO_4^{2-} and whether the ions seemed to be present at a high, moderate, or low concentration. A black background may be useful for these three tests.

9. If some of the results are difficult to discern, place your microplate on an overhead projector. Examine the drops for signs of cloudiness. Looking at the drops from the side, keep your line of vision 10° to 15° above the plane of the lid. Compare each drop tested with the control drops in row 2. If any sign of cloudiness is detected in a test sample, it is due to the Tyndall effect and is a positive test result. Record your results.

DISPOSAL

10. Clean all equipment and your lab station. Return equipment to its proper place. Dispose of chemicals and solutions in the containers designated by your teacher. Do not pour any chemicals down the drain or in the trash unless your teacher directs you to do so. Wash your hands thoroughly before you leave the lab and after all work is finished.

DATA TABLE				
Test for:	Fe^{3+}	Ca^{2+}	Cl^-	SO_4^{2-}
Reacting with:	SCN^-	$C_2O_4^{2-}$	Ag^+	Sr^{2+}
Reference (all four ions)				
Distilled H_2O (control—no ions)				
Tap water				
Bottled spring water				
River or lake water				
Solution X				

Analysis

1. **Organizing Ideas** Describe what each positive test looked like. Write the balanced chemical equations and net ionic equations for each of the positive tests.

_____

_____

_____

_____

_____

_____

_____

_____

_____

_____

_____

_____

Conclusions

1. **Organizing Conclusions** List the solutions that you tested and the ions that you found in each solution. Include notes on whether the concentration of each ion was high, moderate, or low based on your observations.

_____

_____

_____

_____

_____

Testing Water for Ions *continued*

2. Predicting Outcomes Using your test results, predict which water sample would be the "hardest." Explain your reasoning.

_____

_____

_____

_____

_____

_____

_____

Quick Lab)

Household Acids and Bases

Which of the household substances are acids, and which are bases?

MATERIALS

- dishwashing liquid, dishwasher detergent, laundry detergent, laundry stain remover, fabric softener, and bleach

- mayonnaise, baking powder, baking soda, white vinegar, cider vinegar, lemon juice, soft drinks, mineral water, and milk

- fresh red cabbage
- hot plate
- beaker, 500 mL or larger
- beakers, 50 mL
- spatula
- tap water
- tongs

Always wear safety goggles and a lab apron to protect your eyes and clothing. If you get a chemical in your eyes, immediately flush the chemical out at the eyewash station while calling to your teacher. Know the locations of the emergency lab shower and the eyewash station and the procedures for using them.

Do not touch any chemicals. If you get a chemical on your skin or clothing, wash the chemical off at the sink while calling to your teacher. Make sure you carefully read the labels and follow the precautions on all containers of chemicals that you use. If there are no precautions stated on the label, ask your teacher what precautions you should follow. Do not taste any chemicals or items used in the laboratory. Never return leftovers to their original container; take only small amounts to avoid wasting supplies.

Do not heat glassware that is broken, chipped, or cracked. Use tongs or a hot mitt to handle heated glassware and other equipment because hot glassware does not always look hot.

PROCEDURE

Record all your results in the data table provided.

1. To make an acid-base indicator, extract juice from red cabbage. First, cut up some red cabbage and place it in a large beaker. Add enough water so that the beaker is half full. Then, bring the mixture to a boil. Let it cool, and then pour off and save the cabbage juice. This solution is an acid-base indicator.

2. Assemble foods, beverages, and cleaning products to be tested.

3. If the substance being tested is a liquid, pour about 5 mL into a small beaker. If it is a solid, place a small amount into a beaker, and moisten it with about 5 mL of water.

4. Add a drop or two of the red cabbage juice to the solution being tested, and note the color. The solution will turn red if it is acidic and green if it is basic.

| Household Acids and Bases *continued*

DISPOSAL

5. Place all solids in the trash. Pour all liquids down the drain.

Data Table			
Substance	**Color with cabbage juice**	**Substance**	**Color with cabbage juice**
dishwashing liquid		mayonnaise	
dishwashing detergent		baking powder	
laundry detergent		baking soda	
laundry stain remover		white vinegar	
fabric softener		cider vinegar	
bleach		lemon juice	
		soft drink	
		mineral water	
		milk	

DISCUSSION

1. Are the cleaning products acids, bases, or neither?

_____

_____

2. What are acid/base characteristics of foods and beverages?

_____

_____

_____

3. Did you find consumer warning labels on basic or acidic products?

_____

_____

_____

_____

Inquiry

Is It an Acid or a Base?

When scientists uncover a problem they need to solve, they think carefully about the problem and then use their knowledge and experience to develop a plan for solving it. In this experiment, you will be given a set of eight colorless solutions. Four of them are acidic solutions (dilute hydrochloric acid) and four are basic solutions (dilute sodium hydroxide). The concentrations of both the acidic and the basic solutions are 0.1 M, 0.2 M, 0.4 M, and 0.8 M. Phenolphthalein has been added to the acidic solutions.

You will first write a procedure to determine which solutions are acidic and which are basic and then carry out your procedure. You will then develop and carry out another procedure that will allow you to order the acidic and basic solutions from lowest to highest concentration. As you plan your procedures, consider the properties of acids and bases that are discussed in Chapter 14. Predict what will happen to a solution of each type and concentration when you do each test. Then compare your predictions with what actually happens. You will have limited amounts of the unknown solutions to work with, so use them carefully. Ask your teacher what additional supplies (if any) will be available to you.

OBJECTIVES

- **Design** an experiment to solve a chemical problem.

- **Relate** observations of chemical properties to identify unknowns.

- **Infer** a conclusion from experimental data.

- **Apply** acid-base concepts.

MATERIALS

- 24-well microplate or 24 small test tubes

- labeled pipets containing solutions numbered 1–8

- toothpicks

    For other supplies, check with your teacher

Always wear safety goggles and a lab apron to protect your eyes and clothing. If you get a chemical in your eyes, immediately flush the chemical out at the eyewash station while calling to your teacher. Know the locations of the emergency lab shower and the eyewash station and the procedures for using them.

Do not touch any chemicals. If you get a chemical on your skin or clothing, wash the chemical off at the sink while calling to your teacher. Make sure you carefully read the labels and follow the precautions on all containers of chemicals that you use. If there are no precautions stated on the label, ask your teacher what precautions you should follow. Do not taste any chemicals or items used in the

laboratory. Never return leftovers to their original container; take only small amounts to avoid wasting supplies.

Never put broken glass in a regular waste container. Broken glass should be disposed of separately according to your teacher's instructions.

PREPARATION

1. Write the steps you will use to determine which solutions are acids and which solutions are bases.

_____

_____

_____

_____

_____

_____

2. Ask your teacher to approve your plan and give you any additional supplies you will need.

3. Record your data in the data tables provided. In **Data Table 1,** record the numbers of the unknown solutions in the correct columns as you identify them. In **Data Table 2,** record the concentration of each solution as you test it, and then record the concentrations of HCl and NaOH present in the solution.

PROCEDURE

1. Carry out your plan for determining which solutions are acids and which are bases. As you perform your tests, avoid letting the tips of the storage pipets come into contact with other chemicals. Squeeze drops out of the pipets onto the 24-well plate and then use these drops for your tests. Record all observations in the space provided and then record your results in **Data Table 1.**

2. Write your procedure for determining the concentrations of the solutions. Ask your teacher to approve your plan, and request any additional supplies you will need.

_____

_____

_____

_____

_____

_____

Is It an Acid or a Base? *continued*

3. Carry out your procedure for determining the concentrations of the solutions. Record all observations, and record your results in the second data table.

DISPOSAL

4. Clean all apparatus and your lab station. Return equipment to its proper place. Dispose of chemicals and solutions in the containers designated by your teacher. Do not pour any chemicals down the drain or in the trash unless your teacher directs you to do so. Wash your hands thoroughly before you leave the lab and after all work is finished.

OBSERVATIONS

Procedure 1 _____

_____

_____

_____

_____

_____

DATA TABLE 1	
Acids	**Bases**

Procedure 2 _____

_____

_____

_____

_____

_____

Is It an Acid or a Base? *continued*

DATA TABLE 2		
Concentration	HCl	NaOH

Conclusions

1. Analyzing Conclusions List the numbers of the solutions and their concentrations.

_____

_____

_____

_____

_____

_____

_____

_____

_____

_____

Is It an Acid or a Base? *continued*

2. Analyzing Conclusions Describe the test results that led you to identify some solutions as acids and others as bases. Explain how you determined the concentrations of the unknown solutions.

_____

_____

_____

_____

_____

_____

_____

_____

EXTENSIONS

1. Evaluating Methods Compare your results with those of another lab group. Do you think that your teacher gave both groups the same set of solutions? (Is your solution 1 the same as their solution 1, and so on?) Explain your reasoning.

_____

_____

_____

_____

_____

_____

Is It an Acid or a Base? *continued*

2. **Applying Conclusions** Imagine that you are helping to clean out the school's chemical storeroom. You find a spill coming from a large unlabeled reagent bottle filled with a clear liquid. What tests would you do to quickly determine if the substance is acidic or basic?

_____

_____

_____

_____

_____

_____

_____

_____

_____

_____

_____

Testing the pH of Rainwater

Do you have acid precipitation in your area?

MATERIALS

- rainwater
- distilled water
- 500 mL jars
- thin, transparent metric ruler (± 0.1 cm)
- pH test paper: narrow range, ± 0.2–0.3, or pH meter

Always wear safety goggles and a lab apron to protect your eyes and clothing. If you get a chemical in your eyes, immediately flush the chemical out at the eyewash station while calling to your teacher. Know the locations of the emergency lab shower and the eyewash station and the procedures for using them.

PROCEDURE

Record all of your results in a data table.

1. Each time it rains, set out five clean jars to collect the rainwater. If the rain continues for more than 24 hours, put out new containers at the end of each 24-hour period until the rain stops. (The same procedure can be used with snow if the snow is allowed to melt before measurements are taken. You may need to use larger containers if a heavy snowfall is expected.)

2. After the rain stops or at the end of each 24-hour period, use a thin, plastic ruler to measure the depth of the water to the nearest 0.1 cm. Using the pH paper, test the water to determine its pH to the nearest 0.2 to 0.3.

3. Record the following information:

   a. the date and time the collection started

   b. the date and time the collection ended

   c. the location where the collection was made (town and state)

   d. the amount of rainfall in centimeters

   e. the pH of the rainwater

4. Find the average pH of each collection that you have made for each rainfall, and record it in **Data Table 1.**

5. Collect samples on at least five different days. The more samples you collect, the more informative your data will be. Make up additional data sheets to record your data.

6. For comparison, determine the pH of pure water by testing five samples of distilled water with pH paper. Record your results in **Data Table 2,** and then calculate an average pH for distilled water.

Testing the pH of Rainwater *continued*

Jar	Start (date & time)	End (date & time)	Location	Rainfall (cm)	pH
Data Table 1					
1					
2					
3					
4					
5					
					Avg. ____

Data Table 2	
Sample	pH
1	
2	
3	
4	
5	
	Avg. ____

DISCUSSION

1. What is the pH of distilled water?

_____

_____

2. What is the pH of normal rainwater? How do you explain any differences between the pH readings?

_____

_____

_____

_____

Testing the pH of Rainwater *continued*

3. What are the drawbacks of using a ruler to measure the depth of collected water? How could you increase the precision of your measurement?

_____

_____

_____

_____

_____

_____

4. Does the amount of rainfall or the time of day the sample is taken have an effect on its pH? Try to explain any variability among samples.

_____

_____

5. What conclusion can you draw from this investigation? Explain how your data support your conclusion.

_____

_____

_____

Name _____ Class _____ Date _____

Microscale

How Much Calcium Carbonate Is in an Eggshell?

The calcium carbonate content of eggshells can be easily determined by means of an acid/base back-titration. In this experiment, a strong acid will react with calcium carbonate in eggshells. Then, the amount of unreacted acid will be determined by titration with a strong base.

OBJECTIVES

• **Determine** the amount of calcium carbonate present in an eggshell.

• **Relate** experimental titration measurements to a balanced chemical equation.

• **Infer** a conclusion from experimental data.

• **Apply** reaction stoichiometry concepts.

MATERIALS

• 1.00 M HCl

• 1.00 M NaOH

• 10 mL graduated cylinder

• 50 mL micro solution bottle or small Erlenmeyer flask

• 100 mL beaker

• balance

• desiccator (optional)

• distilled water

• drying oven

• eggshell

• forceps

• mortar and pestle

• phenolphthalein solution

• thin-stemmed pipets or medicine droppers, 3

• weighing paper

Always wear safety goggles and a lab apron to protect your eyes and clothing. If you get a chemical in your eyes, immediately flush the chemical out at the eyewash station while calling to your teacher. Know the locations of the emergency lab shower and the eyewash station and the procedures for using them.

Do not touch any chemicals. If you get a chemical on your skin or clothing, wash the chemical off at the sink while calling to your teacher. Make sure you carefully read the labels and follow the precautions on all containers of chemicals that you use. If there are no precautions stated on the label, ask your teacher what precautions you should follow. Do not taste any chemicals or items used in the laboratory. Never return leftovers to their original container; take only small amounts to avoid wasting supplies.

Acids and bases are corrosive. If an acid or base spills onto your skin or clothing, wash the area immediately with running water. Call your teacher in the event of an acid spill. Acid or base spills should be cleaned up promptly.

| How Much Calcium Carbonate Is in an Eggshell? *continued*

PREPARATION

1. Wash an empty eggshell with distilled water and carefully peel all the membranes from its inside. Place *all* of the shell in a premassed beaker and dry the shell in the drying oven at 110°C for about 15 min.

2. Use **Data Table 1** and **Data Table 2** to record your data.

3. Put exactly 5.0 mL of water in the 10.0 mL graduated cylinder. Record this volume in **Data Table 1.** Label the first pipet *Acid*. To calibrate the pipet, fill it with water. **Do not use this pipet for the base solution.** Holding the pipet vertically, add 20 drops of water to the cylinder. Record the new volume of water in the graduated cylinder in **Data Table 1** under Trial 1.

4. Without emptying the graduated cylinder, add an additional 20 drops from the pipet. Record the new volume for Trial 2. Repeat this procedure once more for Trial 3.

5. Repeat Preparation steps **3** and **4** for the second pipet. Label this pipet *Base*. **Do not use this pipet for the acid solution.**

6. Make sure that the three trials produce data that are similar to one another. If one is greatly different from the others, perform Preparation steps **3–5** again.

7. Remove the eggshell and beaker from the oven. Cool them in a desiccator. Record the mass of the entire eggshell in **Data Table 2.** Place half of the shell into the clean mortar, and grind the shell into a very fine powder.

PROCEDURE

1. Measure the mass of a piece of weighing paper. Transfer about 0.1 g of ground eggshell to a piece of weighing paper, and measure the eggshell's mass as accurately as possible. Record the mass in **Data Table 2.** Place this eggshell sample into a clean, 50 mL micro solution bottle (or Erlenmeyer flask).

2. Fill the acid pipet with 1.00 M HCl acid solution, and then empty the pipet into an extra 100 mL beaker. Label the beaker "Waste." Fill the base pipet with the 1.00 M NaOH base solution, and then empty the pipet into the waste beaker.

3. Fill the acid pipet once more with 1.00 M HCl. Holding the acid pipet vertically, add exactly 150 drops of 1.00 M HCl to the bottle or flask that contains the eggshell. Swirl the flask gently for 3 to 4 min. Observe the reaction taking place. Wash down the sides of the flask with about 10 mL of distilled water. Using a third pipet, add two drops of phenolphthalein solution.

4. Fill the base pipet with the 1.00 M NaOH. Slowly add NaOH from the base pipet into the bottle or flask that contains the eggshell reaction mixture. Stop adding base when the mixture remains a faint pink color, even after it is swirled gently. **Be sure to add the base drop by drop, and be certain the drops end up in the reaction mixture and not on the walls of the bottle or flask.** Keep careful count of the number of drops used. Record in **Data Table 2** the number of drops of base used.

How Much Calcium Carbonate Is in an Eggshell? *continued*

DISPOSAL

5. Clean all equipment and your lab station. Dispose of chemicals and solutions as directed by your teacher. Wash your hands thoroughly before you leave the lab.

Data Table 1				
Graduated Cylinder Readings (Pipet Calibration: Steps 3–5)				
Trial	**Initial— acid pipet (mL)**	**Final— acid pipet (mL)**	**Initial— base pipet (mL)**	**Final— base pipet (mL)**
1				
2				
3				

Total volume of drops—acid pipet _____

Average volume of each drop _____

Total volume of drops—base pipet _____

Average volume of each drop _____

Data Table 2	
Mass of entire eggshell (g)	
Mass of ground eggshell sample (g)	
Number of drops of 1.00 M HCl added	
Volume of 1.00 M HCl added (mL)	
Moles of HCl added	
Number of drops of 1.00 M NaOH added	
Volume of 1.00 M NaOH added (mL)	
Moles of NaOH added	
Number of moles of HCl reacted with eggshell	
Number of moles of $CaCO_3$ reacted with HCl	
Mass of $CaCO_3$ in eggshell sample (g)	
% of $CaCO_3$ in eggshell sample	

| **How Much Calcium Carbonate Is in an Eggshell?** *continued*

Analysis

1. **Organizing Ideas:** The calcium carbonate in the eggshell sample undergoes a double-displacement reaction with the HCl in step **3.** Write a balanced chemical equation for this reaction. (Hint: The gas observed was CO_2.)

_____

_____

2. **Organizing Ideas:** Write the balanced chemical equation for the acid/base neutralization of the excess unreacted HCl with the NaOH.

_____

_____

_____

3. **Organizing Data:** Calculate the volume of each drop in milliliters. Then convert the number of drops of HCl into volume in milliliters. Record this volume in **Data Table 2.** Repeat this step for the drops of NaOH.

4. **Organizing Data:** Using the relationship between the molarity and volume of acid and the molarity and volume of base needed to neutralize it, calculate the volume of the HCl solution that was neutralized by the NaOH, and record it in **Data Table 2.** (Hint: This relationship was discussed in Section 2.)

_____

_____

_____

_____

_____

_____

How Much Calcium Carbonate Is in an Eggshell? *continued*

5. Analyzing Results: Calculate the volume and the number of moles of HCl that reacted with the CaCO₃ and record both in **Data Table 2.**

_____

_____

_____

_____

_____

_____

_____

Conclusions

1. Organizing Data: Use the balanced equation for the reaction to calculate the number of moles of CaCO₃ that reacted with the HCl, and record this number in **Data Table 2.**

2. Organizing Data: Use the periodic table to calculate the molar mass of CaCO₃. Then, use the number of moles of CaCO₃ to calculate the mass of CaCO₃ in your eggshell sample. Record this mass in **Data Table 2.** Using the mass of CaCO₃, calculate the percentage of CaCO₃ in your eggshell and record it in **Data Table 2.**

Calorimetry and Hess's Law

Hess's law states that the overall enthalpy change in a reaction is equal to the sum of the enthalpy changes in the individual steps in the process. In this experiment, you will use a calorimeter to measure the energy released in three chemical reactions. From your experimental data, you will verify Hess's law.

OBJECTIVES

- **Demonstrate** proficiency in the use of calorimeters and related equipment.
- **Relate** temperature changes to enthalpy changes.
- **Determine** enthalpies of reaction for several reactions.
- **Demonstrate** that enthalpies of reactions can be additive.

MATERIALS

- 4 g NaOH pellets
- 50 mL 1.0 M HCl acid solution
- 50 mL 1.0 M NaOH solution
- 100 mL 0.50 M HCl solution
- 100 mL graduated cylinder
- balance
- distilled water

- forceps
- glass stirring rod
- gloves
- plastic-foam cups (or calorimeter)
- spatula
- thermometer
- watch glass

Always wear safety goggles and a lab apron to protect your eyes and clothing. If you get a chemical in your eyes, immediately flush the chemical out at the eyewash station while calling to your teacher. Know the locations of the emergency lab shower and the eyewash station and the procedures for using them.

Do not touch any chemicals. If you get a chemical on your skin or clothing, wash the chemical off at the sink while calling to your teacher. Make sure you carefully read the labels and follow the precautions on all containers of chemicals that you use. If there are no precautions stated on the label, ask your teacher what precautions you should follow. Do not taste any chemicals or items used in the laboratory. Never return leftovers to their original container; take only small amounts to avoid wasting supplies.

PREPARATION

1. Use the **Data Table** provided to record the total volumes of liquid, initial temperature, and final temperature of the three reactions you will carry out, as well as the mass of the empty watch glass and the watch glass plus NaOH pellets.

| Calorimetry and Hess's Law *continued*

2. Gently insert the thermometer into the plastic foam cup held upside down. **Thermometers break easily, so be careful with them, and do not use them to stir a solution.**

PROCEDURE

Reaction 1: Dissolving NaOH

1. Pour 100 mL of distilled water into your calorimeter. Record the water temperature to the nearest 0.1°C.

2. Weigh a clean and dry watch glass to the nearest 0.01 g. Wearing gloves and using forceps, place about 2 g of NaOH pellets on the watch glass. Measure and record the mass of the watch glass and the pellets to the nearest 0.01 g. **It is important that this step be done quickly: NaOH absorbs moisture from the air.**

3. Immediately place the NaOH pellets in the calorimeter cup, and gently stir the solution with a stirring rod. **Do not stir with a thermometer.** Place the lid on the calorimeter. Watch the thermometer, and record the highest temperature in the **Data Table.**

4. Be sure to clean all equipment and rinse it with distilled water before continuing.

Reaction 2: NaOH and HCl in Solution

5. Pour 50 mL of 1.0 M HCl into your calorimeter. Record the temperature of the HCl solution to the nearest 0.1°C.

6. Pour 50 mL of 1.0 M NaOH into a graduated cylinder. **For this step only, rinse the thermometer, and measure the temperature of the NaOH solution in the graduated cylinder to the nearest 0.1°C. Record the temperature, and then replace the thermometer in the calorimeter.**

7. Pour the NaOH solution into the calorimeter cup, and stir gently. Place the lid on the calorimeter. Watch the thermometer and record the highest temperature.

8. Pour the solution in the container designated by your teacher. Clean and rinse all equipment before continuing with the procedure.

Reaction 3: Solid NaOH and HCl in Solution

9. Pour 100 mL of 0.50 M HCl into your calorimeter. Record the temperature of the HCl solution to the nearest 0.1°C.

10. Measure the mass of a clean and dry watch glass, and record the mass. Wear gloves, and using forceps, obtain approximately 2 g of NaOH pellets. Place them on the watch glass, and record the total mass. **As in step 2, it is important that this step be done quickly.**

11. Immediately place the NaOH pellets in the calorimeter, and gently stir the solution. Place the lid on the calorimeter. Watch the thermometer, and record the highest temperature. When finished with this reaction, pour the solution into the container designated by your teacher for disposal of mostly neutral solutions.

DISPOSAL

12. Check with your teacher for the proper disposal procedures. Always wash your hands thoroughly after cleaning up the lab area and equipment.

Data Table			
Reaction	**1**	**2**	**3**
Volume of liquid(s), mL			
Initial temperature, °C			
Highest temperature, °C			
Mass of empty watch glass, g			
Mass of watch glass + NaOH, g			

Analysis

1. Organizing Ideas Write a balanced chemical equation for each of the three reactions that you performed. (Hint: Be sure to include the physical states of matter for all substances.)

_____

_____

_____

_____

2. Organizing Ideas Write the equation for the total reaction by adding two of the equations from item **1** and then canceling out substances that appear in the same form on both sides of the new equation.

_____

_____

| Calorimetry and Hess's Law *continued*

3. Organizing Data Calculate the change in temperature for each of the reactions.

_____

_____

_____

_____

4. Organizing Data Assuming that the density of the water and the solutions is 1.00 g/mL, calculate the mass of liquid present for each of the reactions.

5. Analyzing Results Using the calorimeter equation, calculate the energy as heat released by each reaction. (Hint: Use the specific heat of water in your calculations.)

$$c_p, H_2O = 4.184 \text{ J/g} \bullet °C$$

$$\text{heat} = m \times \Delta t \times c_p, H_2O$$

6. Organizing Data Calculate the moles of NaOH used in each of the reactions.

Calorimetry and Hess's Law *continued*

7. Analyzing Results Calculate the ΔH value in kJ/mol of NaOH for each of the three reactions.

8. Organizing Ideas Using your answer to Analysis item **2** and your knowledge of Hess's law, explain how the enthalpies for the three reactions should be mathematically related.

_____

_____

_____

Name _____ Class _____ Date _____

(Quick Lab)

Factors Influencing Reaction Rate

How do the type of reactants, surface area of reactants, concentration of reactants, and catalysts affect the rates of chemical reactions?

MATERIALS

- Bunsen burner
- paper ash
- copper foil strip
- graduated cylinder, 10 mL
- magnesium ribbon
- matches
- paper clip

- sandpaper
- steel wool
- 2 sugar cubes
- white vinegar
- zinc strip
- 6 test tubes, 16 × 150 mm
- tongs

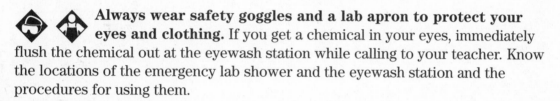

Always wear safety goggles and a lab apron to protect your eyes and clothing. If you get a chemical in your eyes, immediately flush the chemical out at the eyewash station while calling to your teacher. Know the locations of the emergency lab shower and the eyewash station and the procedures for using them.

PROCEDURE

Remove all combustible material from the work area.

1. Add 10 mL of vinegar to each of three test tubes. To one test tube, add a 3 cm piece of magnesium ribbon; to a second, add a 3 cm zinc strip; and to a third, add a 3 cm copper strip. (All metals should be the same width.) If necessary, polish the metals with sandpaper until they are shiny. Record your results in **Data Table 1.**

2. Using tongs, hold a paper clip in the hottest part of the burner flame for 30 s. Repeat with a ball of steel wool 2 cm in diameter. Record your results in **Data Table 2.**

3. To one test tube, add 10 mL of vinegar; to a second, add 5 mL of vinegar plus 5 mL of water; and to a third, add 2.5 mL of vinegar plus 7.5 mL of water. To each of the three test tubes, add a 3 cm piece of magnesium ribbon. Record your results in **Data Table 3.**

4. Using tongs, hold a sugar cube and try to ignite it with a match. Then try to ignite it in a burner flame. Rub paper ash on a second cube, and try to ignite it with a match. Record your results in **Data Table 4.**

DISPOSAL

5. Combine all liquids and pour them down the drain. Save all metal strips for reuse, but if they are too corroded, put them in the trash. Put all other solids in the trash.

Factors Influencing Reaction Rate *continued*

Data Table 1		
10 mL vinegar + Mg ribbon	10 mL vinegar + Zn strip	10 mL vinegar + Cu strip

Data Table 2	
Paper clip in flame	Steel wool in flame

Data Table 3		
10 mL vinegar + Mg ribbon	5 mL vinegar + 5 mL H_2O + Mg ribbon	2.5 mL vinegar + 7.5 mL water + Mg ribbon

Data Table 4		
Sugar cube with match	Sugar cube with burner	Sugar cube + paper ash with match

Factors Influencing Reaction Rate *continued*

DISCUSSION

1. What are the rate-influencing factors in each step of the procedure?

_____

_____

_____

2. What were the results from each step of the procedure? How do you interpret each result?

_____

_____

_____

_____

_____

_____

_____

_____

_____

Name _____ Class _____ Date _____

Rate of a Chemical Reaction

In this experiment, you will determine the rate of the reaction whose net equation is written as follows:

$$3Na_2S_2O_5(aq) + 2KIO_3(aq) + 3H_2O(l) \xrightarrow{H^+} 2KI(aq) + 6NaHSO_4(aq)$$

One way to study the rate of this reaction is to observe how fast $Na_2S_2O_5$ is used up. After all the $Na_2S_2O_5$ solution has reacted, the concentration of iodine, I_2, an intermediate in the reaction, increases. A starch indicator solution, added to the reaction mixture, will change from clear to a blue-black color in the presence of I_2.

In the procedure, the concentrations of the reactants are given in terms of drops of solution A and drops of solution B. Solution A contains $Na_2S_2O_5$, the starch indicator solution, and dilute sulfuric acid to supply the hydrogen ions needed to catalyze the reaction. Solution B contains KIO_3. You will run the reaction with several different concentrations of the reactants and record the time it takes for the blue-black color to appear.

OBJECTIVES

- **Prepare** and **observe** several different reaction mixtures.

- **Demonstrate** proficiency in measuring reaction rates.

- **Relate** experimental results to a rate law that can be used to predict the results of various combinations of reactants.

MATERIALS

- 8-well microscale reaction strips, 2

- distilled or deionized water

- fine-tipped dropper bulbs or small microtip pipets, 3

- solution A

- solution B

- stopwatch or clock with second hand

 Always wear safety goggles and a lab apron to protect your eyes and clothing. If you get a chemical in your eyes, immediately flush the chemical out at the eyewash station while calling to your teacher. Know the locations of the emergency lab shower and the eyewash station and the procedures for using them.

PREPARATION

1. Use the **Data Table** provided to record your data.

2. Obtain three dropper bulbs or small microtip pipets, and label them *A*, *B*, and H_2O.

3. Fill the bulb of pipet A with solution A, the bulb of pipet B with solution B, and the bulb of pipet for H_2O with distilled water.

PROCEDURE

1. Using the first 8-well strip, place five drops of solution A into each of the first five wells. Record the number of drops in the appropriate places in the **Data Table. For the best results, try to make all drops about the same size.**

2. In the second 8-well reaction strip, place one drop of solution B in the first well, two drops in the second well, three drops in the third well, four drops in the fourth well, and five drops in the fifth well. Record the number of drops in the **Data Table.**

3. In the second 8-well strip that contains drops of solution B, add four drops of water to the first well, three drops to the second well, two drops to the third well, and one drop to the fourth well. Do not add any water to the fifth well.

4. Carefully invert the second strip. The surface tension should keep the solutions from falling out of the wells. Place the strip well-to-well on top of the first strip.

5. Hold the strips tightly together and record the exact time, or set the stopwatch, as you shake the strips once. This procedure should mix the upper solutions with each of the corresponding lower ones.

6. Observe the lower wells. Note the sequence in which the solutions react, and record the number of seconds it takes for each solution to turn a blue-black color.

DISPOSAL

7. Dispose of the solutions in the container designated by your teacher. Wash your hands thoroughly after cleaning up the area and equipment.

Data Table					
Well	**1**	**2**	**3**	**4**	**5**
Time rxn. began					
Time rxn. stopped					
Drops of A					
Drops of B					
Drops of H_2O					

Rate of a Chemical Reaction *continued*

Analysis

1. Organizing Data: Calculate the time elapsed for the complete reaction of each combination of solutions A and B.

_____

_____

_____

_____

_____

2. Evaluating Data: Make a graph of your results. Label the x-axis "Number of drops of solution B." Label the y-axis "Time elapsed." Make a similar graph for drops of solution B versus rate (1/time elapsed).

3. Analyzing Information: Which mixture reacted the fastest? Which mixture reacted the slowest?

_____

_____

_____

_____

Rate of a Chemical Reaction *continued*

4. **Evaluating Methods:** Why was it important to add the drops of water to the wells that contained fewer than five drops of solution B? (Hint: Figure out the total number of drops in each of the reaction wells.)

_____

_____

_____

_____

Conclusions

1. **Evaluating Conclusions:** Which of the following variables that can affect the rate of a reaction is tested in this experiment: temperature, catalyst, concentration, surface area, or nature of reactants? Explain your answer.

_____

_____

_____

2. **Applying Ideas:** Use your data and graphs to determine the relationship between the concentration of solution B and the rate of the reaction. Describe this relationship in terms of a rate law.

_____

_____

_____

EXTENSIONS

1. **Predicting Outcomes:** What combination of drops of solutions A and B would you use if you wanted the reaction to last exactly 2.5 min?

_____

_____

_____

_____

_____

Microscale

DATASHEET FOR IN-TEXT LAB

Measuring K_a for Acetic Acid

The acid dissociation constant, K_a, is a measure of the strength of an acid. Strong acids are completely ionized in water. Because weak acids are only partly ionized, they have a characteristic K_a value. Properties that depend on the ability of a substance to ionize, such as conductivity and colligative properties, can be used to measure K_a. In this experiment, you will compare the conductivity of a 1.0 M solution of acetic acid, CH_3COOH, a weak acid, with the conductivities of solutions of varying concentrations of hydrochloric acid, HCl, a strong acid. From the comparisons you make, you will be able to estimate the concentration of hydronium ions in the acetic acid solution and calculate its K_a.

OBJECTIVES

- **Compare** the conductivities of solutions of known and unknown hydronium ion concentrations.

- **Relate** conductivity to the concentration of ions in solution.

- **Explain** the validity of the procedure on the basis of the definitions of strong and weak acids.

- **Compute** the numerical value of K_a for acetic acid.

MATERIALS

- 1.0 M acetic acid, CH_3COOH
- 1.0 M hydrochloric acid, HCl
- 24-well plate
- distilled or deionized water
- LED conductivity testers
- paper towels
- thin-stemmed pipets

Always wear safety goggles and a lab apron to protect your eyes and clothing. If you get a chemical in your eyes, immediately flush the chemical out at the eyewash station while calling to your teacher. Know the locations of the emergency lab shower and the eyewash station and the procedures for using them.

Do not touch any chemicals. If you get a chemical on your skin or clothing, wash the chemical off at the sink while calling to your teacher. Make sure you carefully read the labels and follow the precautions on all containers of chemicals that you use. If there are no precautions stated on the label, ask your teacher what precautions you should follow. Do not taste any chemicals or items used in the laboratory. Never return leftovers to their original container; take only small amounts to avoid wasting supplies.

Acids and bases are corrosive. If an acid or base spills onto your skin or clothing, wash the area immediately with running water. Call your teacher in the event of an acid spill. Acid or base spills should be cleaned up promptly.

| Measuring K_a for Acetic Acid *continued*

PROCEDURE

1. Obtain samples of 1.0 M HCl solution and 1.0 M CH_3COOH solution.

2. Place 20 drops of HCl in one well of a 24-well plate. Place 20 drops of CH_3COOH in an adjacent well. Label the location of each sample.

3. Test the HCl and CH_3COOH with the conductivity tester. Note the relative intensity of the tester light for each solution. After testing, rinse the tester probes with distilled water. Remove any excess moisture with a paper towel.

4. Place 18 drops of distilled water in each of six wells in your 24-well plate. Add 2 drops of 1.0 M HCl to the first well to make a total of 20 drops of solution. Mix the contents of this well thoroughly by picking the contents up in a pipet and returning them to the well.

5. Repeat this procedure by taking 2 drops of the previous dilution and placing it in the next well containing 18 drops of water. Return any unused solution in the pipet to the well from which it was taken. Mix the new solution with a new pipet. (You now have 1.0 M HCl in the well from Procedure step **2,** 0.10 M HCl in the first dilution well, and 0.010 M HCl in the second dilution.)

6. Continue diluting in this manner until you have six successive dilutions. The $[H_3O^+]$ should now range from 1.0 M to 1.0×10^{-6} M. Write the concentrations in the first column of the **Data Table** provided.

7. Using the conductivity tester, test the cells containing HCl in order from most concentrated to least concentrated. Note the brightness of the tester bulb, and compare it with the brightness of the bulb when it was placed in the acetic acid solution. (Retest the acetic acid well any time for comparison.) After each test, rinse the tester probes with distilled water, and use a paper towel to remove any excess moisture. When the brightness produced by one of the HCl solutions is about the same as that produced by the acetic acid, you can infer that the two solutions have about the same hydronium ion concentration and that the pH of the HCl solution is equal to the pH of the acetic acid. If the glow from the bulb is too faint to see, turn off the lights or build a light shield around your conductivity tester bulb.

8. Record the results of your observations by noting which HCl concentration causes the intensity of the bulb to most closely match that of the bulb when it is in acetic acid. (Hint: If the conductivity of no single HCl concentration matches that of the acetic acid, then estimate the value between the two concentrations that match the best.)

DISPOSAL

9. Clean your lab station. Clean all equipment, and return it to its proper place. Dispose of chemicals and solutions in containers designated by your teacher. Do not pour any chemicals down the drain or throw anything in the trash unless your teacher directs you to do so. Wash your hands thoroughly after all work is finished and before you leave the lab.

Measuring K_a for Acetic Acid *continued*

DATA TABLE	
HCl concentration	**Observations and comparisons**

Analysis

1. **Resolving Discrepancies** How did the conductivity of the 1.0 M HCl solution compare with that of the 1.0 M CH_3COOH solution? Why do you think this was so?

_____

_____

_____

_____

_____

_____

2. **Organizing Data** What is the H_3O^+ concentration of the HCl solution that most closely matched the conductivity of the acetic acid?

_____

_____

_____

_____

_____

3. Drawing Conclusions What was the H_3O^+ concentration of the 1.0 M CH_3COOH solution? Why?

_____

_____

_____

_____

_____

Conclusions

1. Applying Models The acid ionization expression for CH_3COOH is the following:

$$K_a = \frac{[H_3O^+]\,[CH_3COO^-]}{[CH_3COOH]}$$

Use your answer to Analysis item **3** to calculate K_a for the acetic acid solution.

2. Applying Models Explain how it is possible for solutions of HCl and CH_3COOH to show the same conductivity but have different concentrations.

_____

_____

_____

_____

_____

_____

_____

_____

Measuring K_a for Acetic Acid continued

EXTENSIONS

1. **Evaluating Methods** Compare the K_a value that you calculated with the value found on page 606 of your text. Calculate the percent error for this experiment.

_____

_____

2. **Predicting Outcomes** Lactic acid ($HOOCCHOHCH_3$) has a K_a of 1.4×10^{-4}. Predict whether a solution of lactic acid would cause the conductivity tester to glow brighter or dimmer than a solution of acetic acid with the same concentration. How noticeable would the difference be?

_____

_____

_____

_____

_____

_____

_____

_____

_____

_____

Quick Lab

Redox Reactions

MATERIALS

- aluminum foil
- beaker, 250 mL
- 1 M copper(II) chloride solution, $CuCl_2$
- 3% hydrogen peroxide
- manganese dioxide

- metric ruler
- scissors
- test-tube clamp
- test tube, 16 × 150 mm
- wooden splint

Always wear safety goggles and a lab apron to protect your eyes and clothing. If you get a chemical in your eyes, immediately flush the chemical out at the eyewash station while calling to your teacher. Know the locations of the emergency lab shower and the eyewash station and the procedures for using them.

Do not touch any chemicals. If you get a chemical on your skin or clothing, wash the chemical off at the sink while calling to your teacher. Make sure you carefully read the labels and follow the precautions on all containers of chemicals that you use. If there are no precautions stated on the label, ask your teacher what precautions you should follow. Do not taste any chemicals or items used in the laboratory. Never return leftovers to their original container; take only small amounts to avoid wasting supplies.

PROCEDURE

Record all of your results in the **Data Table.**

1. Put 10 mL of hydrogen peroxide in a test tube, and add a small amount of manganese dioxide (equal to the size of about half a pea). What is the result?

2. Insert a glowing wooden splint into the test tube. What is the result? If oxygen is produced, a glowing wooden splint inserted into the test tube will glow brighter.

3. Fill the 250 mL beaker halfway with the copper(II) chloride solution.

4. Cut foil into 2 cm × 12 cm strips.

5. Add the aluminum strips to the copper(II) chloride solution. Use a glass rod to stir the mixture, and observe for 12 to 15 minutes. What is the result?

DISPOSAL

6. Check with your teacher for the proper disposal procedures. Always wash your hands thoroughly after cleaning up the lab area and equipment.

| Redox Reactions *continued*

Data Table	
Reactants	**Result**
H_2O_2 + MgO	
H_2O_2 + MgO with glowing splint	
aluminum foil + copper(II) chloride	

DISCUSSION

1. Write balanced equations showing what happened in each of the reactions.

_____

_____

2. Write a conclusion for the two experiments.

_____

_____

_____

_____

_____

Microscale

Reduction of Manganese in Permanganate Ion

In Chapter 15, you studied acid-base titrations in which an unknown amount of acid is titrated with a carefully measured amount of base. In this procedure, a similar approach called a *redox titration* is used. In a redox titration, the reducing agent, Fe^{2+}, is oxidized to Fe^{3+} by the oxidizing agent, MnO_4^-. When this process occurs, the Mn in MnO_4^- changes from a +7 to a +2 oxidation state and has a noticeably different color. You can use this color change to signify a redox reaction "end point." When the reaction is complete, any excess MnO_4^- added to the reaction mixture will give the solution a pink or purple color. The volume data from the titration, the known molarity of the $KMnO_4$ solution, and the mole ratio from the balanced redox equation will give you the information you need to calculate the molarity of the $FeSO_4$ solution.

OBJECTIVES

- **Demonstrate** proficiency in performing redox titrations and recognizing end points of a redox reaction.

- **Write** a balanced oxidation-reduction equation for a redox reaction.

- **Determine** the concentration of a solution by using stoichiometry and volume data from a titration.

MATERIALS

- 0.0200 M $KMnO_4$
- 1.0 M H_2SO_4
- 100 mL graduated cylinder
- 125 mL Erlenmeyer flasks, 4
- 250 mL beakers, 2
- 400 mL beaker

- burets, 2
- distilled water
- double buret clamp
- $FeSO_4$ solution
- ring stand
- wash bottle

Always wear safety goggles and a lab apron to protect your eyes and clothing. If you get a chemical in your eyes, immediately flush the chemical out at the eyewash station while calling to your teacher. Know the locations of the emergency lab shower and the eyewash station and the procedures for using them.

Do not touch any chemicals. If you get a chemical on your skin or clothing, wash the chemical off at the sink while calling to your teacher. Make sure you carefully read the labels and follow the precautions on all containers of chemicals that you use. If there are no precautions stated on the label, ask your teacher what precautions you should follow. Do not taste any chemicals or items used in the laboratory. Never return leftovers to their original container; take only small amounts to avoid wasting supplies.

Reduction of Manganese in Permanganate Ion *continued*

 Acids and bases are corrosive. If an acid or base spills onto your skin or clothing, wash the area immediately with running water. Call your teacher in the event of an acid spill. Acid or base spills should be cleaned up promptly.

Never put broken glass in a regular waste container. Broken glass should be disposed of separately according to your teacher's instructions.

PREPARATION

1. Use the data table provided to record your data.

2. Clean two 50 mL burets with a buret brush and distilled water. Rinse each buret at least three times with distilled water to remove contaminants.

3. Label one 250 mL beaker *0.0200 M KMnO₄* and the other *FeSO₄*. Label three of the flasks *1*, *2*, and *3*. Label the 400 mL beaker *Waste*. Label one buret *KMnO₄* and the other *FeSO₄*.

4. Measure approximately 75 mL of 0.0200 M KMnO₄, and pour it into the appropriately labeled beaker. Obtain approximately 75 mL of FeSO₄ solution, and pour it into the appropriately labeled beaker.

5. Rinse one buret three times with a few milliliters of 0.0200 M KMnO₄ from the appropriately labeled beaker. Collect these rinses in the waste beaker. Rinse the other buret three times with small amounts of FeSO₄ solution from the appropriately labeled beaker. Collect these rinses in the waste beaker.

6. Set up the burets as instructed by your teacher. Fill one buret with approximately 50 mL of 0.0200 M KMnO₄ from the beaker, and fill the other buret with approximately 50 mL of the FeSO₄ solution from the other beaker.

7. With the waste beaker underneath its tip, open the KMnO₄ buret long enough to be sure the buret tip is filled. Repeat the process for the FeSO₄ buret.

8. Add 50 mL of distilled water to one of the 125 mL Erlenmeyer flasks, and add one drop of 0.0200 M KMnO₄ to the flask. Set this mixture aside to use as a color standard. It can be compared with the titration mixture to determine the end point.

PROCEDURE

1. Record in the **Data Table** the initial buret readings for both solutions. Add 10 mL of the hydrated iron(II) sulfate solution, FeSO₄•7H₂O, to the flask labeled *1*. Add 5 mL of 1 M H₂SO₄ to the FeSO₄ solution in this flask. The acid will help keep the Fe²⁺ ions in the reduced state, which will allow you time to titrate.

2. Slowly add KMnO₄ from the buret to the FeSO₄ in the flask while swirling the flask. When the color of the solution matches the color standard you prepared in Preparation step **8,** record in the **Data Table** the final readings of the burets.

3. Empty the titration flask into the waste beaker. Repeat the titration procedure in steps **1** and **2** with the flasks labeled *2* and *3*.

DISPOSAL

4. Dispose of the contents of the waste beaker in the container designated by your teacher. Also, pour the color-standard flask into this container. Wash your hands thoroughly after cleaning up the area and equipment.

Data Table

Trial	Initial KMnO$_4$ volume (mL)	Final KMnO$_4$ volume (mL)	Initial FeSO$_4$ volume (mL)	Final FeSO$_4$ volume (mL)
1				
2				
3				

Analysis

1. Organizing Ideas: Write the balanced equation for the redox reaction of FeSO$_4$ and KMnO$_4$.

_____

_____

2. Evaluating Data: Calculate the number of moles of MnO$_4^-$ reduced in each trial.

Reduction of Manganese in Permanganate Ion *continued*

3. Analyzing Information: Calculate the number of moles of Fe^{2+} oxidized in each trial.

4. Applying Conclusions: Calculate the average concentration (molarity) of the iron(II) sulfate solution.

Reduction of Manganese in Permanganate Ion *continued*

EXTENSIONS

1. Designing Experiments: What possible sources of error can you identify with this procedure? If you can think of ways to eliminate them, ask your teacher to approve your plan, and run the procedure again.

_____

_____

_____

_____

Voltaic Cells

In voltaic cells, oxidation and reduction half-reactions take place in separate half-cells, which can consist of a metal electrode immersed in a solution of its metal ions. The electrical potential, or voltage, that develops between the electrodes is a measure of the combined reducing strength of one reactant and oxidizing strength of the other reactant.

OBJECTIVES

- **Construct** a Cu-Zn voltaic cell.

- **Design** and construct two other voltaic cells.

- **Measure** the potential of the voltaic cells.

- **Evaluate** cells by comparing the measured cell voltages with the voltages calculated from standard reduction potentials.

MATERIALS

- 0.5 M $Al_2(SO_4)_3$, 75 mL
- 0.5 M $CuSO_4$, 75 mL
- 0.5 M $ZnSO_4$, 75 mL
- Aluminum strip, 1 cm × 8 cm
- Copper strip, 1 cm × 8 cm
- Zinc strip, 1 cm × 8 cm
- Distilled water

- 100 mL graduated cylinder
- Emery cloth
- 150 mL beakers, 3
- Salt bridge
- Voltmeter
- Wires with alligator clips, 2

Always wear safety goggles and a lab apron to protect your eyes and clothing. If you get a chemical in your eyes, immediately flush the chemical out at the eyewash station while calling to your teacher. Know the locations of the emergency lab shower and the eyewash station and the procedures for using them.

Do not touch any chemicals. If you get a chemical on your skin or clothing, wash the chemical off at the sink while calling to your teacher. Make sure you carefully read the labels and follow the precautions on all containers of chemicals that you use. If there are no precautions stated on the label, ask your teacher what precautions you should follow. Do not taste any chemicals or items used in the laboratory. Never return leftovers to their original container; take only small amounts to avoid wasting supplies.

PREPARATION

1. Use the **Data Table** provided to record your data.

| Voltaic Cells *continued*

2. Remove any oxide coating from strips of aluminum, copper, and zinc by rubbing them with an emery cloth. Keep the metal strips dry until you are ready to use them.

3. Label three 150 mL beakers $Al_2(SO_4)_3$, $CuSO_4$, and $ZnSO_4$.

PROCEDURE

1. Pour 75 mL of 0.5 M $ZnSO_4$ into the $ZnSO_4$ beaker and 75 mL of 0.5 M $CuSO_4$ into the $CuSO_4$ beaker.

2. Place one end of the salt bridge into the $CuSO_4$ solution and the other end into the $ZnSO_4$ solution.

3. Place a zinc strip into the zinc solution and a copper strip into the copper solution.

4. Using the alligator clips, connect one wire to one end of the zinc strip and the second wire to the copper strip. Take the free end of the wire attached to the zinc strip, and connect it to one terminal on the voltmeter. Take the free end of the wire attached to the copper strip, and connect it to the other terminal on the voltmeter. The needle on the voltmeter should move to the right. If your voltmeter's needle points to the left, reverse the way the wires are connected to the terminals of the voltmeter. Immediately record the voltage reading in the **Data Table,** and disconnect the circuit.

5. Record the concentration of the solutions and sketch a diagram of your electrochemical cell.

6. Rinse the copper and zinc strips with a *very small* amount of distilled water. Collect the rinse from the copper strip in the $CuSO_4$ beaker and the rinse from the zinc strip in the $ZnSO_4$ beaker. Rinse each end of the salt bridge into the corresponding beaker.

7. Use the table of standard reduction potentials in the textbook to calculate the standard voltages for the other cells you can build using copper, zinc, or aluminum. Build these cells and measure their potentials following steps **1–6.**

DISPOSAL

8. Clean all apparatus and your lab station. Wash your hands. Place the pieces of metal in the containers designated by your teacher. Each solution should be poured in its own separate disposal container. Do not mix the contents of the beakers.

Voltaic Cells *continued*

Data Table

Cell	$Zn|Zn^{2+}||Cu^{2+}|Cu$	$Al|Al^{3+}||Zn^{2+}|Zn$	$Al|Al^{3+}||Cu^{2+}|Cu$
Diagram			
Conc.			
Voltage (V)			

Analysis

1. **Organizing Ideas** For each cell that you constructed, write the equations for the two half-cell reactions. Obtain the standard half-cell potentials for the half-reactions from the table in the textbook, and write these E^O values after the equations.

2. **Organizing Ideas** For each cell you tested, combine the two half-reactions to obtain the equation for the net reaction.

_____

_____

_____

_____

Voltaic Cells *continued*

3. **Organizing Ideas** Use the E^O values for the half-reactions to determine the E^O for each cell.

_____

_____

_____

4. **Resolving Discrepancies** Compare the actual cell voltages you measured with the standard cell voltages in item 3. Explain why you would expect a difference.

_____

_____

_____

Conclusions

1. **Inferring Conclusions** Based on the voltages that you measured, which cell produces the most energy?

_____

_____

2. **Applying Ideas** On the basis of your data, which metal is the strongest reducing agent? Which metal ion is the strongest oxidizing agent?

_____

_____

_____

3. **Applying Ideas** Indicate the direction of electron flow in each of your cell diagrams.

Voltaic Cells *continued*

EXTENSIONS

1. **Predicting Outcomes** Describe how and why the reactions would stop if the cells had been left connected.

_____

_____

_____

_____

2. **Designing Experiments** Design a method that could use several of the electrochemical cells you constructed to generate more voltage than any individual cell provided. (Hint: consider what would happen if you linked an Al-Zn cell and a Zn-Cu cell. If your teacher approves your plan, test your idea.)

_____

_____

_____

_____

_____

Microscale

Simulation of Nuclear Decay Using Pennies and Paper

Radioactive isotopes are unstable. All radioactive matter decays, or breaks down, in a predictable pattern. Radioactive isotopes release radiation as they disintegrate into daughter isotopes.

   The rate of decay is a measure of how fast an isotope changes into its daughter isotope. The rate of radioactive decay is conveniently characterized by the isotope's half-life, the period of time it takes one-half of the original material to decay. Half-lives vary from billions of years to fractions of a second.

OBJECTIVES

• **Infer** that the rate of decay can be simulated by a random process.

• **Compare** the numbers of pennies that are showing heads with the number showing tails.

• **Create** a string plot that represents nuclear decay.

• **Relate** observations to the rate of nuclear decay.

• **Graph** the data.

• **Compare** the results of the two simulation procedures.

MATERIALS

• colored paper or cloth strips, approximately 65 cm × 2.5 cm (2 strips)

• graph paper

• one sheet of stiff poster board, 70 cm × 60 cm

• pennies or other objects supplied by your teacher (100)

• scissors, tape, meter stick, pencil, and string

• shoe box with lid

Always wear safety goggles and a lab apron to protect your eyes and clothing. If you get a chemical in your eyes, immediately flush the chemical out at the eyewash station while calling to your teacher. Know the locations of the emergency lab shower and the eyewash station and the procedures for using them.

PREPARATION

For Part A, record your data in the **Data Table.**

PROCEDURE

Part A: Simulating radioactive decay with pennies

1. Place 100 pennies into the shoe box so that the head sides are up. The pennies will represent atoms. Record 100 in the "Unchanged atoms" column and 0 in the "Changed atoms" column.

Simulation of Nuclear Decay Using Pennies and Paper *continued*

2. With the lid on the box, shake the box up and down 5 times. We will count each shaking period as being equivalent to 10 s.

3. Open the lid, and remove all of the pennies that have the tails side up. These pennies represent the changed atoms.

4. Count the number of pennies remaining in the box. Record this number in the 10 s row of the "Unchanged atoms" column in the **Data Table.** Count the number of changed atoms (the pennies that you removed from the box), and record the number in the 10 s row.

5. Each lab partner should predict how many times steps **2–4** will need to be repeated until only one unchanged atom remains. Record the time that each lab partner predicted. Remember that each shaking period is counted as 10 s, so four shaking periods would be recorded as 40 s.

6. Repeat steps **2–4** by counting and recording each time until only 1 (or 0) penny with the head side up remains.

Part B: Simulating decay with paper

7. Draw an y-axis and x-axis on the poster board so that they are about 5 cm from the left side and the bottom edge respectively. Label the x-axis as "Time" and the y-axis as "Amount of material."

8. Along the x-axis, draw marks every 10 cm from the y-axis line. Label the first mark "0" and the next mark "1," and so on. Each mark represents 1 minute.

9. Place one of the colored strips vertically with its lower edge centered on the 0 mark of the x-axis. Tape the strip in place.

10. Fold the other colored strip in half, and cut it in the middle. Place one-half of the strip so that it is centered on the next mark, and tape the strip in place.

11. Fold the remaining piece of the strip in half, and cut it exactly in the middle.

12. Place one of the pieces so that it is centered on the next mark, and tape the piece in place.

13. Repeat steps **11** and **12,** and each time, tape the first piece vertically at the next x-axis mark. Continue until you have at least 8 strips taped along the x-axis.

14. Use the string to join the tops of each strip of paper to make a continuous curve.

DISPOSAL

15. Return the pennies and box to your teacher. Dispose of the poster board, strips, and string as instructed by your teacher. Clean up your lab station.

Simulation of Nuclear Decay Using Pennies and Paper *continued*

Data Table		
Time (s)	Unchanged atoms	Changed atoms
0	100	0
10		
20		
30		
40		
50		
60		
70		

Analysis

PART A

1. Predicting Outcomes How long did it take to have only 1 penny (0 pennies) left in the box? How close was your prediction in step 5?

_____

_____

2. Analyzing Data Make a graph of your data on a piece of graph paper. Label the x-axis "Time" and the y-axis "Unchanged atoms." Plot the number of unchanged atoms versus time. Draw a smooth curve through the data points.

Simulation of Nuclear Decay Using Pennies and Paper *continued*

3. Analyzing Results Each trial was comparable to a 10 s period of time. How long did it take for half of your pennies to be removed from the box? What is the half-life of the process?

_____

_____

4. Interpreting Graphics Use your graph to determine the time it takes to have only 25% of the unchanged atoms remaining. In your experiment, how many pennies remained in the box at that time?

_____

_____

_____

PART B

5. Analyzing Results How many half lives have passed after 4 minutes?

_____

_____

6. Interpreting Graphics Using the string plot, determine how many minutes it took until only 20% of the original material remained.

_____

_____

Conclusions

1. Inferring Conclusions If you started with a paper strip that was twice as long, would the half-life change?

_____

_____

2. Inferring Conclusions Is there a relationship between the graph from Part A and the string plot from Part B?

_____

_____

_____

Polymers and Toy Balls

What polymers make the best toy balls? Two possibilities are latex rubber and a polymer produced from ethanol and sodium silicate. Latex rubber is a polymer of covalently bonded atoms.

The polymer formed from ethanol, C_2H_5OH, and a solution of sodium silicate, $Na_2Si_3O_7$, also has covalent bonds. It is known as *water glass* because it dissolves in water.

In this experiment, you will synthesize rubber and the ethanol silicate polymer and test their properties.

OBJECTIVES

- **Synthesize** two different polymers.

- **Prepare** a small toy ball from each polymer.

- **Observe** the similarities and differences between the two types of balls.

- **Measure** the density of each polymer.

- **Compare** the bounce height of the two balls.

MATERIALS

- 2 L beaker, or plastic bucket or tub
- 3 mL 50% ethanol solution
- 5 oz paper cups, 2
- 10 mL 5% acetic acid solution (vinegar)
- 25 mL graduated cylinder
- 10 mL graduated cylinder

- 10 mL liquid latex
- 12 mL sodium silicate solution
- distilled water
- gloves
- meterstick
- paper towels
- wooden stick

Always wear safety goggles and a lab apron to protect your eyes and clothing. If you get a chemical in your eyes, immediately flush the chemical out at the eyewash station while calling to your teacher. Know the locations of the emergency lab shower and the eyewash station and the procedures for using them.

Do not touch any chemicals. If you get a chemical on your skin or clothing, wash the chemical off at the sink while calling to your teacher. Make sure you carefully read the labels and follow the precautions on all containers of chemicals that you use. If there are no precautions stated on the label, ask your teacher what precautions you should follow. Do not taste any chemicals or items used in the laboratory. Never return leftovers to their original container; take only small amounts to avoid wasting supplies.

Polymers and Toy Balls *continued*

PREPARATION

1. **Organizing Data** Use **Data Table 1** and **Data Table 2** to record your observations.

PROCEDURE

1. Fill the 2 L beaker, bucket, or tub about half-full with distilled water.

2. Using a clean 25 mL graduated cylinder, measure 10 mL of liquid latex and pour it into one of the paper cups.

3. Thoroughly clean the 25 mL graduated cylinder with soap and water, and then rinse it with distilled water.

4. Measure 10 mL of distilled water. Pour it into the paper cup with the latex.

5. Measure 10 mL of the 5% acetic acid solution, and pour it into the paper cup with the latex and water.

6. Immediately stir the mixture with the wooden stick.

7. As you continue stirring, a polymer lump will form around the wooden stick. Pull the stick with the polymer lump from the paper cup, and immerse the lump in the 2 L beaker, bucket, or tub.

8. While wearing gloves, gently pull the lump from the wooden stick. Be sure to keep the lump immersed under the water.

9. Keep the latex rubber underwater, and use your gloved hands to mold the lump into a ball. Then, squeeze the lump several times to remove any unused chemicals. You may remove the latex rubber from the water as you roll it in your hands to smooth the ball.

10. Set aside the latex rubber ball to dry. While it is drying, proceed to step **11.**

11. In a clean 25 mL graduated cylinder, measure 12 mL of sodium silicate solution, and pour it into the other paper cup.

12. In a clean 10 mL graduated cylinder, measure 3 mL of 50% ethanol. Pour the ethanol into the paper cup with the sodium silicate, and mix with the wooden stick until a solid substance is formed.

13. While wearing gloves, remove the polymer that forms and place it in the palm of one hand. Gently press it with the palms of both hands until a ball that does not crumble is formed. This step takes a little time and patience. The liquid that comes out of the ball is a combination of ethanol and water. Occasionally, moisten the ball by letting a small amount of water from a faucet run over it. When the ball no longer crumbles, you are ready to go to the next step.

14. Observe as many physical properties of the balls as possible, and record your observations in **Data Table 1** or **Data Table 2.**

15. Drop each ball several times, and record your observations.

Polymers and Toy Balls *continued*

16. Drop one ball from a height of 1 m, and measure its bounce. Perform three trials for each ball.

17. Measure the diameter and mass of each ball. Record these data in **Data Table 1** or **Data Table 2.**

DISPOSAL

18. Dispose of any extra solutions in the containers indicated by your teacher. Clean up your lab area. Remember to wash your hands thoroughly when your lab work is finished.

Data Table 1	
Latex rubber	
Bounce height—trial 1	
Bounce height—trial 2	
Bounce height—trial 3	
Mass	
Diameter	
Other observations	

Data Table 2	
Ethanol-silicate polymer	
Bounce height—trial 1	
Bounce height—trial 2	
Bounce height—trial 3	
Mass	
Diameter	
Other observations	

Polymers and Toy Balls *continued*

ANALYSIS

1. **Analyzing Information** List at least three of your observations of the properties of the two balls.

   _____

   _____

   _____

   _____

   _____

2. **Organizing Data** Calculate the average height of the bounce for each type of ball.

3. **Organizing Data** Calculate the volume for each ball. Even though the balls may not be perfectly spherical, assume that they are. (Hint: The volume of a sphere is equal to $\frac{4}{3} \times \pi \times r^3$, where r is the radius of the sphere, which is one-half of the diameter.) Then, calculate the density of each ball, using your mass measurements.

Polymers and Toy Balls *continued*

CONCLUSIONS

1. **Inferring Conclusions** Which polymer would you recommend to a toy company for making new toy balls? Explain your reasoning.

_____

_____

_____

_____

2. **Evaluating Viewpoints** What are some other possible practical applications for each of the polymers you made?

_____

_____

EXTENSION

1. **Predicting Outcomes** Explain why you would not be able to measure the volumes of the balls by submerging them in water.

_____

_____

_____

_____

(Skills Practice)
Casein Glue

Cow's milk contains averages of 4.4% fat, 3.8% protein, and 4.9% lactose. At the normal pH of milk, 6.3 to 6.6, the protein remains dispersed because it has a net negative charge due to the dissociation of the carboxylic acid group, as shown in Figure A below. As the pH is lowered by the addition of an acid, the protein acquires a net charge of zero, as shown in Figure B. After the protein loses its negative charge, it can no longer remain in solution, and it coagulates into an insoluble mass. The precipitated protein is known as casein and has a molecular mass between 75 000 and 375 000 amu. The pH at which the net charge on a protein becomes zero is called the isoelectric pH. For casein, the isoelectric pH is 4.6.

$$H_2N— \boxed{protein} —COO^- \quad {}^+H_3N— \boxed{protein} —COO^-$$

Figure A   **Figure B**

In this experiment, you will coagulate the protein in milk by adding acetic acid. The casein can then be separated from the remaining solution by filtration. This process is known as separating the curds from the whey. The excess acid in the curds can be neutralized by the addition of sodium hydrogen carbonate, $NaHCO_3$. The product of this reaction is casein glue. Do not eat or drink any materials or products of this lab.

OBJECTIVES

• **Recognize** the structure of a protein.

• **Predict** and **observe** the result of acidifying milk.

• **Prepare** and **test** a casein glue.

• **Deduce** the charge distribution in proteins as determined by pH.

MATERIALS

• 100 mL graduated cylinder
• 250 mL beaker
• 250 mL Erlenmeyer flask
• funnel
• glass stirring rod
• hot plate
• medicine dropper

• baking soda, $NaHCO_3$
• nonfat milk
• paper
• paper towel
• thermometer
• white vinegar
• wooden splints, 2

Casein Glue *continued*

Always wear safety goggles and a lab apron to protect your eyes and clothing. If you get a chemical in your eyes, immediately flush the chemical out at the eyewash station while calling to your teacher. Know the locations of the emergency lab shower and the eyewash station and the procedures for using them.

Do not touch any chemicals. If you get a chemical on your skin or clothing, wash the chemical off at the sink while calling to your teacher. Make sure you carefully read the labels and follow the precautions on all containers of chemicals that you use. If there are no precautions stated on the label, ask your teacher what precautions you should follow. Do not taste any chemicals or items used in the laboratory. Never return leftovers to their original container; take only small amounts to avoid wasting supplies.

PREPARATION

1. Use the **Data Table** provided for recording observations at each step of the procedure.

2. Predict the characteristics of the product that will be formed when the acetic acid is added to the milk. Record your predictions in the **Data Table.**

PROCEDURE

1. Pour 125 mL of nonfat milk into a 250 mL beaker. Add 20 mL of 4% acetic acid (white vinegar).

2. Place the mixture on a hot plate and heat it to 60°C. Record your observations in the **Data Table** and compare them with the predictions you made in Preparation step **2.**

3. Filter the mixture through a folded piece of paper towel into an Erlenmeyer flask.

4. Discard the filtrate, which contains the whey. Scrape the curds from the paper towel back into the 250 mL beaker.

5. Add 1.2 g of $NaHCO_3$ to the beaker and stir. Slowly add drops of water, stirring intermittently, until the consistency of white glue is obtained.

6. Use your glue to fasten together two pieces of paper. Also fasten together two wooden splints. Allow the splints to dry overnight, and then test the joint for strength.

DISPOSAL

7. Clean all apparatus and your lab station. Return equipment to its proper place. Dispose of chemicals and solutions in the containers designated by your teacher. Do not pour any chemicals down the drain or in the trash unless your teacher directs you to do so. Wash your hands thoroughly before you leave the lab and after all work is finished.

Casein Glue *continued*

Data Table	
Predicted Result: milk + acetic acid	
Actual Result milk: + acetic acid	

Analysis

1. Organizing Ideas Write the net ionic equation for the reaction between the excess acetic acid and the sodium hydrogen carbonate. Include the physical states of the reactants and products.

_____

_____

2. Evaluating Methods In this experiment, what happened to the lactose and fat portions of the milk?

_____

_____

Conclusion

1. Inferring Conclusions **Figure A** shows that the net charge on a protein is negative at pH values higher than its isoelectric pH because the carboxyl group is ionized. **Figure B** shows that at the isoelectric pH, the net charge is zero. Predict the net charge on a protein at pH values lower than the isoelectric point, and draw a diagram to represent the protein.

_____

_____

_____

Extensions

1. Relating Ideas **Figure B** represents a protein as a dipolar ion, or zwitterion. The charges in a zwitterion suggest that the carboxyl group donates a hydrogen ion to the amine group. Is there any other way to represent the protein in **Figure B** so that it still has a net charge of zero?

_____

_____

Casein Glue *continued*

2. **Designing Experiments** Design a strength-testing device for the glue joint between the two wooden splints. If your teacher approves your design, create the device and use it to test the strength of the glue.

_____

_____

_____

Extraction and Filtration

Extraction, the separation of substances in a mixture by using a solvent, depends on solubility. For example, sand can be separated from salt by adding water to the mixture. The salt dissolves in the water, and the sand settles to the bottom of the container. The sand can be recovered by decanting the water. The salt can then be recovered by evaporating the water.

Filtration separates substances based on differences in their physical states or in the size of their particles. For example, a liquid can be separated from a solid by pouring the mixture through a paper-lined funnel or, if the solid is more dense than the liquid, the solid will settle to the bottom of the container, leaving the liquid on top. The liquid can then be decanted, leaving the solid.

SETTLING AND DECANTING

1. Fill an appropriate-sized beaker with the solid-liquid mixture provided by your teacher. Allow the beaker to sit until the bottom is covered with solid particles and the liquid is clear.

2. Grasp the beaker with one hand. With the other hand, pick up a stirring rod and hold it along the lip of the beaker. Tilt the beaker slightly so that liquid begins to pour out in a slow, steady stream.

GRAVITY FILTRATION

1. Prepare a piece of filter paper. Fold it in half and then in half again. Tear the corner of the filter paper, and open the filter paper into a cone. Place it in the funnel.

2. Put the funnel, stem first, into a filtration flask, or suspend it over a beaker using an iron ring.

3. Wet the filter paper with distilled water from a wash bottle. The paper should adhere to the sides of the funnel, and the torn corner should prevent air pockets from forming between the paper and the funnel.

4. Pour the mixture to be filtered down a stirring rod into the filter. The stirring rod directs the mixture into the funnel and reduces splashing.

5. Do not let the level of the mixture in the funnel rise above the edge of the filter paper.

6. Use a wash bottle to rinse all of the mixture from the beaker into the funnel.

VACUUM FILTRATION

1. Check the T attachment to the faucet. Turn on the water. Water should run without overflowing the sink or spitting while creating a vacuum. To test for a vacuum, cover the opening of the horizontal arm of the T with your thumb or index finger. If you feel your thumb being pulled inward, you have a vacuum. Note the number of turns of the knob that are needed to produce the flow of water that creates a vacuum.

2. Turn the water off. Attach the pressurized rubber tubing to the *horizontal* arm of the T. (You do not want water to run through the tubing.)

3. Attach the free end of the rubber tubing to the side arm of a filter flask. Check for a vacuum. Turn on the water so that it rushes out of the faucet (refer to step 1). Place the palm of your hand over the opening of the **Erlenmeyer** flask. You should feel the vacuum pull your hand inward. If you do not feel any pull or if the pull is weak, increase the flow of water. If increasing the flow of water fails to work, shut off the water and make sure your tubing connections are tight.

4. Insert the neck of a Büchner funnel into a one-hole rubber stopper until the stopper is about two-thirds to three-fourths up the neck of the funnel. Place the funnel stem into the **Erlenmeyer** flask so that the stopper rests in the mouth of the flask.

5. Obtain a piece of round filter paper. Place it inside the Büchner funnel over the holes. Turn on the water as in step 1. Hold the filter flask with one hand, place the palm of your hand over the mouth of the funnel, and check for a vacuum.

6. Pour the mixture to be filtered into the funnel. Use a wash bottle to rinse all of the mixture from the beaker into the funnel.

(Skills Practice)
Gravimetric Analysis

Gravimetric analytical methods are based on accurate and precise mass measurements. They are used to determine the amount or percentage of a compound or element in a sample material. For example, if we want to determine the percentage of iron in an ore or the percentage of chloride ion in drinking water, gravimetric analysis would be used.

A gravimetric procedure generally involves reacting the sample to produce a reaction product that can be used to calculate the mass of the element or compound in the original sample. For example, to calculate the percentage of iron in a sample of iron ore, the mass of the ore is determined. The ore is then dissolved in hydrochloric acid to produce $FeCl_3$. The $FeCl_3$ precipitate is converted to a hydrated form of Fe_2O_3 by adding water and ammonia to the system. The mixture is then filtered to separate the hydrated Fe_2O_3 from the mixture. The hydrated Fe_2O_3 is heated in a crucible to drive the water from the hydrate, producing anhydrous Fe_2O_3. The mass of the crucible and its contents is determined after successive heating steps to ensure that the product has reached constant mass and that all of the water has been driven off. The mass of Fe_2O_3 produced can be used to calculate the mass and percentage of iron in the original ore sample.

Gravimetric procedures require accurate and precise techniques and measurements to obtain suitable results. Possible sources of error are the following:

1. The product (precipitate) that is formed is contaminated.

2. Some product is lost when transferring the product from a filter to a crucible.

3. The empty crucible is not clean or is not at constant mass for the initial mass measurement.

4. The system is not heated sufficiently to obtain an anhydrous product.

Always wear safety goggles and a lab apron to protect your eyes and clothing. If you get a chemical in your eyes, immediately flush the chemical out at the eyewash station while calling to your teacher. Know the locations of the emergency lab shower and the eyewash station and the procedures for using them.

When using a Bunsen burner, confine long hair and loose clothing. Do not heat glassware that is broken, chipped, or cracked. Use tongs or a hot mitt to handle heated glassware and other equipment; heated glassware does not always look hot. If your clothing catches fire, WALK to the emergency lab shower and use it to put out the fire.

Never put broken glass or ceramics in a regular waste container. Broken glass and ceramics should be disposed of in a separate container designated by your teacher.

Gravimetric Analysis *continued*

SETTING UP THE EQUIPMENT

1. Attach a metal ring clamp to a ring stand, and lay a clay triangle on the ring.

CLEANING THE CRUCIBLE

2. Wash and dry a metal or ceramic crucible and lid. Cover the crucible with its lid, and use a balance to obtain its mass. If the balance is located far from your working station, use crucible tongs to place the crucible and lid on a piece of wire gauze. Carry the crucible to the balance, using the wire gauze as a tray.

HEATING THE CRUCIBLE TO OBTAIN A CONSTANT MASS

3. After recording the mass of the crucible and lid, suspend the crucible over a Bunsen burner by placing it on the clay triangle. Then place the lid on the crucible so that the entire contents are covered.

4. Light the Bunsen burner. Heat the crucible for 5 minutes with a gentle flame, and then adjust the burner to produce a strong flame. Heat for 5 minutes more. Shut off the gas to the burner. Allow the crucible and lid to cool. Using crucible tongs, carry the crucible and lid to the balance. Measure and record the mass. If the mass differs from the mass before heating, repeat the process until mass data from heating trials are within 1% of each other. This assumes that the crucible has a constant mass. The crucible is now ready to be used in a gravimetric analysis procedure. Details will be found in the following experiments. Gravimetric methods are used in Experiment 7 to synthesize magnesium oxide, and to separate $SrCO_3$ from a solution in Experiment 9.

Skills Practice

Paper Chromatography

Chromatography is a technique used to separate substances dissolved in a mixture. The Latin roots of the word are *chromato*, which means "color," and *graphy*, which means "to write." Paper is one medium used to separate the components of a solution.

Paper is made of cellulose fibers that are pressed together. As a solution passes over the fibers and through the pores, the paper acts as a filter and separates the mixture's components. Particles of the same component group together, producing a colored band. Properties such as particle size, molecular mass, and charge of the different solute particles in the mixture affect the distance the components will travel on the paper. The components of the mixture that are the most soluble in the solvent and the least attracted to the paper will travel the farthest. Their band of color will be closest to the edge of the paper.

Always wear safety goggles and a lab apron to protect your eyes and clothing. If you get a chemical in your eyes, immediately flush the chemical out at the eyewash station while calling to your teacher. Know the location of the emergency lab shower and the eyewash station and the procedures for using them.

PROCEDURE

1. Use a lead pencil to sketch a circle about the size of a quarter in the center of a piece of circular filter paper that is 12 cm in diameter.

2. Write one numeral for each substance, including any unknowns, around the inside of this circle. In this experiment, 6 mixtures are to be separated, so the circle is labeled 1 through 6, as shown in **Figure A.**

3. Use a micropipet to place a spot of each substance to be separated next to a number. Make one spot per number. If the spot is too large, you will get a broad, tailing trace with little or no detectable separation.

4. Use the pencil to poke a small hole in the center of the spotted filter paper. Insert a wick through the hole. A wick can be made by rolling a triangular piece of filter paper into a cylinder: start at the point of the triangle, and roll toward its base.

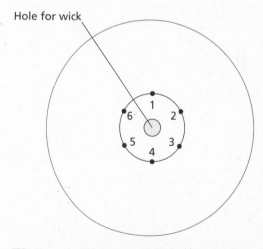

Figure A Filter paper used in paper chromatography is spotted with the mixtures to be separated. Each spot is labeled with a numeral or a name that identifies the mixture to be separated. A hole punched in the center of the paper will attach to a wick that delivers the solvent to the paper.

Paper Chromatography *continued*

5. Fill a petri dish or lid two-thirds full of solvent (usually water or alcohol).

6. Set the bottom of the wick in the solvent so that the filter paper rests on the top of the petri dish.

7. When the solvent is 1 cm from the outside edge of the paper, remove the paper from the petri dish, and allow the chromatogram to dry.

Most writing or drawing inks are mixtures of various components that give them specific properties. Therefore, paper chromatography can be used to study the composition of an ink. Experiment 12 investigates the composition of ball-point pen ink.

(Skills Practice)

Volumetric Analysis

Volumetric analysis, the quantitative determination of the concentration of a solution, is achieved by adding a substance of known concentration to a substance of unknown concentration until the reaction between them is complete. The most common application of volumetric analysis is titration.

A buret is used in titrations. The solution with the known concentration is usually in the buret. The solution with the unknown concentration is usually in the Erlenmeyer flask. A few drops of a visual indicator also are added to the flask. The solution in the buret is then added to the flask until the indicator changes color, signaling that the reaction between the two solutions is complete. Then, using the volumetric data obtained and the balanced chemical equation for the reaction, the unknown concentration is calculated.

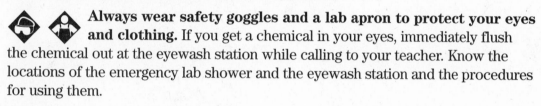 **Always wear safety goggles and a lab apron to protect your eyes and clothing.** If you get a chemical in your eyes, immediately flush the chemical out at the eyewash station while calling to your teacher. Know the locations of the emergency lab shower and the eyewash station and the procedures for using them.

ASSEMBLING THE APPARATUS

1. Attach a buret clamp to a ring stand.

2. Thoroughly wash and rinse a buret. If water droplets cling to the walls of the buret, wash it again and gently scrub the inside walls with a buret brush.

3. Attach the buret to one side of the buret clamp.

4. Place a Erlenmeyer flask for waste solutions under the buret tip.

OPERATING THE STOPCOCK

1. The stopcock should be operated with the left hand. This method gives better control but may prove awkward at first for right-handed students. The handle should be moved with the thumb and first two fingers of the left hand.

2. Rotate the stopcock back and forth. It should move freely and easily. If it sticks or will not move, ask your teacher for assistance. Turn the stopcock to the closed position. Use a wash bottle to add 10 mL of distilled water to the buret. Rotate the stopcock to the open position. The water should come out in a steady stream. If no water comes out or if the stream of water is blocked, ask your teacher to check the stopcock for clogs.

FILLING THE BURET

1. To fill the buret, place a funnel in the top of the buret. Slowly and carefully pour the solution of known concentration from a beaker into the funnel. Open the stopcock, and allow some of the solution to drain into the waste beaker. Then add enough solution to the buret to raise the level above the zero mark, but do not allow the solution to overflow.

READING THE BURET

1. Drain the buret until the bottom of the meniscus is on the zero mark or within the calibrated portion of the buret. If the solution level is not at zero, record the exact reading. If you start from the zero mark, your final buret reading will equal the amount of solution added. Remember, burets can be read to the second decimal place. Burets are designed to read the volume of liquid delivered to the flask, so numbers increase as you read downward from the top. .

2. Replace the waste beaker with an Erlenmeyer flask containing a measured amount of the solution of unknown concentration.

Experiment 15-1 is an example of a back-titration applied to an acid-base reaction; it can be performed on a larger scale if micropipets are replaced with burets.

Calorimetry

Calorimetry, the measurement of the transfer of energy as heat, allows chemists to determine thermal constants, such as the specific heat of metals and the enthalpy of solution.

When two substances at different temperatures touch one another, energy as heat flows from the warmer substance to the cooler substance until the two substances are at the same temperature. The amount of energy transferred is measured in joules. (One joule equals 4.184 calories.)

A device used to measure the transfer of energy as heat is a calorimeter. Calorimeters vary in construction depending on the purpose and the accuracy of the energy measurement required. No calorimeter is a perfect insulator; some energy is always lost to the surroundings as heat. Therefore, every calorimeter must be calibrated to obtain its calorimeter constant.

Always wear safety goggles and a lab apron to protect your eyes and clothing. If you get a chemical in your eyes, immediately flush the chemical out at the eyewash station while calling to your teacher. Know the locations of the emergency lab shower and the eyewash station and the procedures for using them.

Turn off hot plates and other heat sources when not in use. Do not touch a hot plate after it has just been turned off; it is probably hotter than you think. Use tongs when handling heated containers. Never hold or touch containers with your hands while heating them.

The steps for constructing a calorimeter made from plastic foam cups follow.

CONSTRUCTING THE CALORIMETER

1. Trim the lip of one plastic foam cup, and use that cup as the top of your calorimeter. The other cup will be used as the base.

2. Use the pointed end of a pencil to gently make a hole in the center of the calorimeter top. The hole should be large enough to insert a thermometer. Make a hole for the wire stirrer. This hole should be positioned so that the wire stirrer can be raised and lowered without interfering with the thermometer.

3. Place the calorimeter in a beaker to prevent it from tipping over.

CALIBRATING A PLASTIC FOAM CUP CALORIMETER

1. Measure 50 mL of distilled water in a graduated cylinder. Pour it into the calorimeter. Measure and record the temperature of the water in the polystyrene cup.

2. Pour another 50 mL of distilled water into a beaker. Set the beaker on a hot plate, and warm the water to about 60°C. Measure and record the temperature of the water.

| Calorimetry *continued*

3. Immediately pour the warm water into the cup. Cover the cup, and move the stirrer gently up and down to mix the contents thoroughly. **Take care not to break the thermometer.**

4. Watch the thermometer and record the highest temperature attained (usually after about 30 s).

5. Empty the calorimeter.

6. The derivation of the equation to find the calorimeter constant starts with the following relationship.

Energy lost by the warm water = Energy gained by the cool water + Energy gained by the calorimeter

$$q_{warm}\ H_2O = q_{cool}\ H_2O + q_{calorimeter}$$

The energy lost as heat by the warm water is calculated by

$$q_{warm\ H_2O} = mass_{warm\ H_2O} \times 4.184\ J/(g \bullet °C) \times \Delta t$$

The energy gained as heat by the calorimeter system equals the energy lost as heat by the warm water. You can use the following equation to calculate the calorimeter constant C′ for your calorimeter.

$$q_{calorimeter} = q_{warm}\ H_2O = (mass_{cool\ H_2O})\ (4.184\ J/g \times °C)\ (\Delta t_{cool\ H_2O}) + \\ C′\ (\Delta t_{cool\ H_2O})$$

Substitute the data from your calibration and solve for C′.

Modern Chemistry   **304**   Pre-Laboratory Procedure